MEASURING AND SUSTAINING THE NEW ECONOMY

THE TELECOMMUNICATIONS CHALLENGE

CHANGING TECHNOLOGIES AND EVOLVING POLICIES

Report of a Symposium

Committee on The Telecommunications Challenge:
Changing Technologies and Evolving Policies

Committee on Measuring and Sustaining the New Economy

Board on Science, Technology, and Economic Policy

Policy and Global Affairs

Charles W. Wessner, Editor

NATIONAL RESEARCH COUNCIL
OF THE NATIONAL ACADEMIES

THE NATIONAL ACADEMIES PRESS
Washington, D.C.
www.nap.edu

THE NATIONAL ACADEMIES PRESS   500 Fifth Street, N.W.   Washington, DC 20001

NOTICE: The project that is the subject of this report was approved by the Governing Board of the National Research Council, whose members are drawn from the councils of the National Academy of Sciences, the National Academy of Engineering, and the Institute of Medicine. The members of the committee responsible for the report were chosen for their special competences and with regard for appropriate balance.

This study was supported by: Contract/Grant No. CMRC-50SBNB9C1080 between the National Academy of Sciences and the U.S. Department of Commerce; Contract/Grant No. NASW-99037, Task Order 103, between the National Academy of Sciences and the National Aeronautics and Space Administration; Contract/Grant No. CMRC-SB134105C0038 between the National Academy of Sciences and the U.S. Department of Commerce; OFED-13416 between the National Academy of Sciences and Sandia National Laboratories; Contract/Grant No. N00014-00-G-0230, DO #23, between the National Academy of Sciences and the Department of the Navy; Contract/Grant No. NSF-EIA-0119063 between the National Academy of Sciences and the National Science Foundation; and Contract/Grant No. DOE-DE-FG02-01ER30315 between the National Academy of Sciences and the U.S. Department of Energy. Additional support was provided by Intel Corporation. Any opinions, findings, conclusions, or recommendations expressed in this publication are those of the author(s) and do not necessarily reflect the views of the organizations or agencies that provided support for the project.

International Standard Book Number 0-309-10087-9 (Book)
International Standard Book Number 0-309-65628-1 (PDF)

Limited copies are available from Board on Science, Technology, and Economic Policy, National Research Council, 500 Fifth Street, N.W., W547, Washington, DC 20001; 202-334-2200.

Additional copies of this report are available from the National Academies Press, 500 Fifth Street, N.W., Lockbox 285, Washington, DC 20055; (800) 624-6242 or (202) 334-3313 (in the Washington metropolitan area); Internet, http://www.nap.edu

THE NATIONAL ACADEMIES
Advisers to the Nation on Science, Engineering, and Medicine

The **National Academy of Sciences** is a private, nonprofit, self-perpetuating society of distinguished scholars engaged in scientific and engineering research, dedicated to the furtherance of science and technology and to their use for the general welfare. Upon the authority of the charter granted to it by the Congress in 1863, the Academy has a mandate that requires it to advise the federal government on scientific and technical matters. Dr. Ralph J. Cicerone is president of the National Academy of Sciences.

The **National Academy of Engineering** was established in 1964, under the charter of the National Academy of Sciences, as a parallel organization of outstanding engineers. It is autonomous in its administration and in the selection of its members, sharing with the National Academy of Sciences the responsibility for advising the federal government. The National Academy of Engineering also sponsors engineering programs aimed at meeting national needs, encourages education and research, and recognizes the superior achievements of engineers. Dr. Wm. A. Wulf is president of the National Academy of Engineering.

The **Institute of Medicine** was established in 1970 by the National Academy of Sciences to secure the services of eminent members of appropriate professions in the examination of policy matters pertaining to the health of the public. The Institute acts under the responsibility given to the National Academy of Sciences by its congressional charter to be an adviser to the federal government and, upon its own initiative, to identify issues of medical care, research, and education. Dr. Harvey V. Fineberg is president of the Institute of Medicine.

The **National Research Council** was organized by the National Academy of Sciences in 1916 to associate the broad community of science and technology with the Academy's purposes of furthering knowledge and advising the federal government. Functioning in accordance with general policies determined by the Academy, the Council has become the principal operating agency of both the National Academy of Sciences and the National Academy of Engineering in providing services to the government, the public, and the scientific and engineering communities. The Council is administered jointly by both Academies and the Institute of Medicine. Dr. Ralph J. Cicerone and Dr. Wm. A. Wulf are chair and vice chair, respectively, of the National Research Council.

www.national-academies.org

Committee on The Telecommunications Challenge: Changing Technologies and Evolving Policies*

William J. Raduchel, *Chair*
Chairman and CEO
Ruckus Network

Mark E. Doms
Senior Economist
Federal Reserve Bank of
  San Francisco

Kenneth Flamm
Dean Rusk Chair in International
  Affairs
LBJ School of Public Affairs
University of Texas at Austin

Dale W. Jorgenson
Samuel W. Morris University Professor
Harvard University

Cherry A. Murray
Deputy Director for Science and
  Technology
Lawrence Livermore National
  Laboratory

Mark B. Myers
Visiting Executive Professor of
  Management
The Wharton School
University of Pennsylvania

Michael R. Nelson
Director of Internet Technology and
  Strategy
International Business Machines

William J. Spencer
Chairman Emeritus, *retired*
International SEMATECH

*As of November 2004.

Committee on Measuring and Sustaining the New Economy*

Project Staff*

Charles W. Wessner
Study Director

Sujai J. Shivakumar
Program Officer

Ken Jacobson
Consultant

McAlister T. Clabaugh
Program Associate

David E. Dierksheide
Program Associate

*As of November 2004.

For the National Research Council (NRC), this project was overseen by the Board on Science, Technology, and Economic Policy (STEP), a standing board of the NRC established by the National Academies of Sciences and Engineering and the Institute of Medicine in 1991. The mandate of the STEP Board is to integrate understanding of scientific, technological, and economic elements in the formulation of national policies to promote the economic well-being of the United States. A distinctive characteristic of STEP's approach is its frequent interactions with public and private-sector decision makers. STEP bridges the disciplines of business management, engineering, economics, and the social sciences to bring diverse expertise to bear on pressing public policy questions. The members of the STEP Board* and the NRC staff are listed below:

*As of November 2004.

Jack Schuler
Chairman
Ventana Medical Systems, Inc.

Alan Wm. Wolff
Managing Partner
Dewey Ballantine

STEP Staff*

Stephen A. Merrill
Executive Director

Charles W. Wessner
Program Director

Craig M. Schultz
Research Associate

Sujai J. Shivakumar
Program Officer

McAlister T. Clabaugh
Program Associate

David E. Dierksheide
Program Associate

---

*As of November 2004.

Contents

Preface

    Significant and sustained increases in semiconductor productivity, predicted by Moore's Law, has ushered a revolution in communications, computing, and information management.[1] This technological revolution is linked to a distinct rise in the mid-1990s of the long-term growth trajectory of the United States.[2] Indeed, U.S. productivity growth has accelerated in recent years, despite a series of negative economic shocks. Analysis by Dale Jorgenson, Mun Ho, and Kevin Stiroh of the sources of this growth over the 1996 to 2003 period suggests that the production and use of information technology account for a large share of the gains. The authors go further to project that during the next decade, private sector productivity growth will continue at a rate of 2.6 percent per year.[3] The "New

---

[1] This is especially so for the computer hardware sector and perhaps for the Internet as well, although there is insufficient empirical evidence on the degree to which the Internet may be responsible. For a discussion of the impact of the Internet on economic growth see *The Economist*, "A Thinker's Guide," March 30, 2000. For a broad study of investment in technology-capital and its use in various sectors, see McKinsey Global Institute, *U.S. Productivity Growth 1995–2000: Understanding the Contribution of Information Technology Relative to Other Factors,* Washington, D.C.: McKinsey & Co., October 2001.

[2] See Dale W. Jorgenson and Kevin J. Stiroh, "Raising the Speed Limit: U.S. Economic Growth in the Information Age" in National Research Council, *Measuring and Sustaining the New Economy: Report of a Workshop,* Dale W. Jorgenson and Charles W. Wessner, eds., Washington, D.C.: National Academy Press, 2002.

[3] Dale W. Jorgenson, Mun S. Ho, and Kevin J. Stiroh, "Will the U.S. Productivity Resurgence Continue?" *FRBNY Current Issues in Economics and Finance*, 10(1), 2004.

Economy" is, thus, not a fad, but a long-term productivity shift of major significance.[4]

The idea of a New Economy brings together the technological innovations, structural changes, and public policy challenges associated with measuring and sustaining this remarkable economic phenomenon.

- Technological innovation—more accurately, the rapid rate of technological innovation in information technology (including computers, software, and telecommunications) and the rapid growth of the Internet—are now widely seen as underpinning the productivity gains that characterize the New Economy.[5] These productivity gains derive from greater efficiencies in the production of computers from expanded use of information technologies.[6] Many therefore believe that the productivity growth of the New Economy draws from the technological innovations found in information technology industries.[7]

- Structural changes arise from a reconfiguration of knowledge networks and business patterns made possible by innovations in information technology. Phenomena, such as business-to-business e-commerce and Internet retailing, are altering how firms and individuals interact, enabling greater efficiency in purchases, production processes, and inventory management.[8] Offshore outsourcing

---

[4]The introduction of advanced productivity-enhancing technologies obviously does not eliminate the business cycle. See Organisation for Economic Co-operation and Development, *Is There a New Economy? A First Report on the OECD Growth Project.* Paris, France: Organisation for Economic Co-operation and Development, 2000, p. 17. See also M. N. Baily and R. Z. Lawrence, "Do We Have an E-conomy?" NBER Working Paper 8243, April 23, 2001, accessed at *<http://www.nber.org/papers/w8243>*.

[5]Broader academic and policy recognition of the New Economy can be seen, for example from the "Roundtable on the New Economy and Growth in the United States" at the 2003 annual meetings of the American Economic Association, held in Washington, D.C. Roundtable participants included Martin Baily, Martin Feldstein, Robert J. Gordon, Dale Jorgenson, Joseph Stiglitz, and Lawrence Summers. Even those who were initially skeptical about the New Economy phenomenon now find that the facts support the belief that faster productivity growth has proved more durable and has spread to other areas of the economy—e.g., retail, banking. See *The Economist*, "The New 'New Economy'," September 11, 2003.

[6]See, for example, Stephen Oliner and Daniel Sichel, "The Resurgence of Growth in the late 1990s: Is Information Technology the Story?" *Journal of Economic Perspectives,* 14(4), 2000. Oliner and Sichel estimate that improvements in the computer industry's own productive processes account for about a quarter of the overall productivity increase. They also note that the use of information technology by all sorts of companies accounts for nearly half the rise in productivity.

[7]See Alan Greenspan's remarks before the White House Conference on the New Economy, Washington D.C., April 5, 2000, accessed at *<www.federalreserve.gov/BOARDDOCS/SPEECHES/2000/20000405.HTM>*. For a historical perspective, see the Proceedings section of this volume. Ken Flamm compares the economic impact of semiconductors today with impact of railroads in the nineteenth century.

[8]See, for example, Brookes Martin and Zaki Wahhaj, "The Shocking Economic Impact of B2B," *Global Economic Paper*, 37, Goldman Sachs, February 3, 2000.

of service production is another manifestation of structural changes made possible by new information and communications technologies. These structural changes are still emerging as the use and applications of the Internet continue to evolve.

• Public policy plays a major role at several levels. This includes the government's role in fostering rules of interaction within the Internet[9] and its discretion in setting and enforcing the rules by which technology firms, among others, compete.[10] More familiarly, public policy concerns particular fiscal and regulatory choices that can affect the rate and focus of investments in sectors such as telecommunications. The government also plays a critical role within the innovation system.[11] It provides national research capacities,[12] incentives to promote education and training in critical disciplines, and funds most of the nation's basic research.[13] The government also plays a major role in stimulating innovation, most broadly through the patent system.[14] Government procurement and awards also encourage the development of new technologies to fulfill national missions

---

[9]Dr. Vinton Cerf notes that the ability of individuals to interact in potentially useful ways within the infrastructure of the still expanding Internet rests on its basic rule architecture: "The reason it can function is that all the networks use the same set of protocols. An important point is these networks are run by different administrations, which must collaborate both technically and economically on a global scale." See comments by Dr. Cerf in National Research Council, *Measuring and Sustaining the New Economy: Report of a Workshop, op cit.* Also in the same volume, see the presentation by Dr. Shane Greenstein on the evolution of the Internet from academic and government-related applications to the commercial world.

[10]The relevance of competition policy to the New Economy is manifested by the intensity of interest in the antitrust case, *United States versus Microsoft*, and associated policy issues.

[11]See Richard Nelson, ed., *National Innovation Systems*, New York: Oxford University Press, 1993.

[12]The STEP Board has recently completed a major review of the role and operation of government-industry partnerships for the development of new technologies. See National Research Council, *Government-Industry Partnerships for the Development of New Technologies: Summary Report*, Charles W. Wessner, ed., Washington, D.C.: The National Academies Press, 2003.

[13]National Research Council, *Trends in Federal Support of Research and Graduate Education*, Stephen A. Merrill, ed., Washington, D.C.: National Academy Press, 2001.

[14]In addition to government-funded research, intellectual property protection plays an essential role in the continued development of the biotechnology industry. See Wesley M. Cohen and John Walsh, "Public Research, Patents and Implications for Industrial R&D in the Drug, Biotechnology, Semiconductor and Computer Industries" in *Government-Industry Partnerships in Biotechnology and Information Technologies: New Needs and New Opportunities*, Charles W. Wessner, ed., Washington, D.C.: National Academy Press, 2002. There is a similar situation in Information Technology with respect to the combination of generally non-appropriable government-originated innovation and appropriable industry intellectual property creation. The economic rationale for government investment is based on the non-appropriablity of many significant information technology innovations, including the most widely used idiomatic data structures and algorithms, as well as design and architectural patterns. In addition, the IT industry relies on a number of technical and process commonalities or standards such as the suite of Internet protocols, programming languages, core design patterns, and architectural styles.

in defense, health, and the environment.[15] Collectively, these public policies play a central role in the development of the New Economy.

The New Economy offers new policy challenges. Modern information and communications technologies make the globalization of research, development and manufacture possible at scales that are unprecedented. This reality has prompted some analysts to argue that information and communication technology and software production are not commodities that the United States can potentially afford to give up overseas suppliers but are an essential part of the economy's production function. They believe that a loss of U.S. leadership in information and communication technology and software will damage the nation's future ability to compete in diverse industries, not least the information and communication technology industry. Given the pervasiveness of semiconductor-based technologies, collateral consequences of a failure to develop adequate policies to sustain national leadership in information and communication technology is likely to extend to a wide variety of sectors from financial services and health care to automobiles, with critical implications for our nation's security and the wellbeing of our citizens. Understanding and responding to this policy challenge requires a multidisciplinary approach to the interconnections among science, technology, and economic policy.

THE CONTEXT OF THIS REPORT

Since 1991 the National Research Council's Board on Science, Technology, and Economic Policy (STEP) has undertaken a program of activities to improve policymakers' understanding of the interconnections among science, technology, and economic policy and their importance to the American economy and its international competitive position. The Board's interest in the New Economy and its underpinnings derive directly from its mandate.

This mandate has previously been reflected in STEP's widely cited volume, *U.S. Industry in 2000,* which assesses the determinants of competitive performance in a wide range of manufacturing and service industries, including those

---

[15]For example, government support played a critical role in the early development of computers. See Kenneth Flamm, *Creating the Computer*, Washington, D.C.: The Brookings Institution, 1988. For an overview of government industry collaboration, see the introduction to the recent report on the Advanced Technology Program, National Research Council, *The Advanced Technology Program: Assessing Outcomes*, Charles W. Wessner, ed., Washington, D.C.: National Academy Press, 2001. The historical and technical case for government-funded research in IT is well documented in reports by the Computer Science and Telecommunications Board (CSTB) of the National Research Council. In particular, see National Research Council, *Innovation in Information Technology,* Washington, D.C.: The National Academies Press, 2003. This volume provides an update of the of the "tire tracks" diagram first published in CSTB's 1995 Brooks-Sutherland report, which depicts the critical role that government funded university research has played in the development of multi-billion-dollar IT industry.

relating to information technology.[16] The Board also undertook a major study, chaired by Gordon Moore of Intel, on how government-industry partnerships can support growth enhancing technologies.[17] Reflecting a growing recognition of the importance of the surge in productivity since 1995, the Board launched a multifaceted assessment, exploring the sources of growth, measurement challenges, and the policy framework required to sustain the New Economy. The first exploratory volume was published in 2002.[18] Subsequent workshops and ensuing reports in this series include *Productivity and Cyclicality in the Semiconductor Industry, Deconstructing the Computer,* and *Software, Growth, and the Future of the U.S. Economy.* The present report, *The Telecommunications Challenge,* examines the importance of telecommunications to the continued expansion in U.S. productivity growth and related policy issues needed to sustain the benefits of the New Economy.

SYMPOSIUM AND DISCUSSIONS

Believing that increased productivity in the semiconductor, computer component, and software and telecommunications industries plays a key role in sustaining the New Economy, the Committee on Measuring and Sustaining the New Economy, under the auspices of the STEP Board, convened a symposium November 15, 2004 at the National Academy of Sciences, Washington, D.C. The symposium on *The Telecommunications Challenge* drew together expertise from leading academics, national accountants, and innovators in the information technology sector (Appendix B lists these individuals). A major purpose of this symposium was to draw together expert knowledge to inform the Committee, which will issue its findings and recommendations on measuring and sustaining the New Economy in a final consensus report of this series.[19]

The "Proceedings" chapter of this volume contains summaries of their workshop presentations and discussions. Given the quality and the number of presentations, summarizing the workshop proceedings has been a challenge. We have made every effort to capture the main points made during the presentations and the ensuing discussions. We apologize for any inadvertent errors or omissions in our summary of the proceedings. The lessons from this symposium and others

---

[16] National Research Council, *U.S. Industry in 2000: Studies in Competitive Performance,* David C. Mowery, ed., Washington, D.C.: National Academy Press, 1999.

[17] For a summary of this multi-volume study, see National Research, *Government-Industry Partnerships for the Development of New Technologies, Summary Report, op. cit.*

[18] National Research Council, *Measuring and Sustaining the New Economy: Report of a Workshop, op. cit.*

[19] National Research Council, *Enhancing Productivity Growth in the Information Age: Measuring and Sustaining the New Economy,* Dale W. Jorgenson and Charles W. Wessner, eds., Washington, D.C.: The National Academies Press, forthcoming.

in this series will contribute to the Committee's final consensus report on *Measuring and Sustaining the New Economy*.

ACKNOWLEDGMENTS

There is considerable interest in the policy community in developing a better understanding of the technological drivers and appropriate regulatory framework for the New Economy, as well as in a better grasp of its operation. This interest is reflected in the support on the part of agencies that have played a role in the creation and development of the New Economy. We are grateful for the participation and the contributions of the National Aeronautical and Space Administration, the Department of Energy, the National Institute of Standards and Technology, the National Science Foundation, and Sandia National Laboratories.

We are indebted to Ken Jacobson for his preparation of the meeting summary. Several members of the STEP staff also deserve recognition for their contributions to the preparation of this report. We wish to thank Sujai Shivakumar for his contributions to the introduction to the report. We are also indebted to McAlister Clabaugh and David Dierksheide for their role in preparing the conference and getting this report ready for publication.

NRC REVIEW

This report has been reviewed in draft form by individuals chosen for their diverse perspectives and technical expertise, in accordance with procedures approved by the National Academies' Report Review Committee. The purpose of this independent review is to provide candid and critical comments that will assist the institution in making its published report as sound as possible and to ensure that the report meets institutional standards for quality and objectivity. The review comments and draft manuscript remain confidential to protect the integrity of the process.

I wish to thank the following individuals for their review of this report: Jaison Abel, Analysis Group; David Clark, Massachusetts Institute of Technology; Shane Greenstein, Northwestern University; Robert Sparks, California Medical Association Foundation; and William Taylor, NERA Economic Consulting.

Although the reviewers listed above have provided many constructive comments and suggestions, they were not asked to endorse the content of the report, nor did they see the final draft before its release. The review of this report was overseen by the National Academies, which was responsible for making certain that an independent examination of this report was carried out in accordance with institutional procedures and that all review comments were carefully considered. Responsibility for the final content of this report rests entirely with the authoring committee and the institution.

STRUCTURE

This report has three parts: an Introduction; a summary of the proceedings of the November 15, 2004 symposium; and finally, a bibliography that provides additional references.

This report represents an important step in a major research effort by the Board on Science, Technology, and Economic Policy to advance our understanding of the factors shaping the New Economy, the metrics necessary to understand it better, and the policies best suited to sustaining the greater productivity and prosperity that it promises.

Dale W. Jorgenson

List of Acronyms

BEA   Bureau of Economic Analysis of the Department of Commerce
CLEC   Competitive Local Exchange Carriers: a telephone company that
         competes with an incumbent local exchange carrier (ILEC) such as a
         Regional Bell Operating Company (RBOC), GTE, ALLNET, etc.
DBS   Direct Broadcast Satellite; describes small-dish, digital satellite systems
         such as DirecTV and Dish Network
DSL   Digital Subscriber Loop; refers to a family of digital telecommunica-
         tions protocols designed to allow high speed data communication over
         the existing copper telephone lines between end-users and telephone
         companies
FCC   Federal Communications Commission
GNP   Gross National Product
ISP   Internet Service Provider
IT   Information Technology
ITU   International Telecommunications Union
IPTV   Internet Protocol Television; a common denominator for systems where
         television and/or video signals are distributed to subscribers or viewers
         using a broadband connection over Internet Protocol

LLU   Local Loop Unbundling is the process of allowing telecommunications operators to use the twisted-pair telephone connections from the telephone exchange's central office to the customer premises. This local loop is owned by the incumbent local exchange carrier.

RBOC   Regional Bell Operating Companies

SETI   Search for Extraterrestrial Intelligence. The SETI institute is dedicated to explore, understand, and explain the origin, nature, and prevalence of life in the universe.

STEP   The Board on Science, Technology, and Economic Policy of the National Academies

TCP/IP Transmission Control Protocol/Internet Protocol; a protocol for communication between computers, used as a standard for transmitting data over networks and as the basis for standard Internet protocols

UBE   Unbundled Network Elements are a requirement mandated by the Telecommunications Act of 1996. They are the parts of the network that the ILECs are required to offer on an unbundled basis. Together, these parts make up a loop that connects to a DSLAM (**D**igital **S**ubscriber **L**ine **A**ccess **M**ultiplexeror), a voice switch, or both. The loop allows non-facilities-based telecommunications providers to deliver service without laying network infrastructure (copper/fiber).

VoIP   Voice over Internet Protocol; this refers to a category of hardware and software that enables people to use the Internet as the transmission medium for telephone calls by sending voice data in packets rather than by traditional circuit transmissions.

WiFi   Wireless Fidelity; a term for certain types of wireless local area networks that use specifications conforming to standards set by the Institute of Electrical and Electronics Engineers

WiMAX   Worldwide Interoperability for Microwave Access is a certification mark for products that pass conformity and interoperability tests for standards set by the Institute of Electrical and Electronics Engineers concerning point-to-multipoint broadband wireless access.

3G   Third Generation; usually used in reference to the next generation digital mobile network

I

INTRODUCTION

Telecommunications in the New Economy

The New Economy refers to a fundamental transformation in the United States economy as businesses and individuals capitalize on new technologies, new opportunities, and national investments in computing, information, and communications technologies. Use of this term reflects a growing conviction that widespread use of these technologies has made possible a sustained rise in the growth trajectory of the U.S. economy.[1]

To understand this phenomenon better, the Board on Science, Technology, and Economic Policy (STEP) of the National Academies has held since 2000 a series of symposia on Measuring and Sustaining the New Economy. These symposia have examined key issues related to semiconductors (the base technology driving the pace of technological development) as well as computers, software, and telecommunications. Taken together, these meetings have produced a comprehensive picture of what is known about the drivers of the New Economy.

---

[1]In the context of this analysis, the New Economy does not refer to the boom economy of the late 1990s. The term is used in this context to describe the acceleration in U.S. productivity growth that emerged in the mid-1990s, in part as a result of the acceleration of Moore's Law and the resulting expansion in the application of loser cost, higher performance information technologies. See Dale W. Jorgenson, Kevin J. Stiroh, Robert J. Gordon, Daniel E. Sichel, "Raising the Speed Limit: U.S. Economic Growth in the Information Age," *Brookings Papers on Economic Activity*, 2000(1):125–235.

This knowledge can help develop policies needed to sustain the benefits of the New Economy.

New telecommunications technologies—the subject of STEP's fifth conference—have contributed significantly to the New Economy. These contributions include the advantages of new product capabilities for businesses and consumers as well as new, more efficient forms of industrial organization made possible by cheaper and more versatile communications. Thus, while the telecom sector accounts, by various measures, for about one percent of the U.S. economy, it is estimated to be responsible for generating about ten percent of the nation's economic growth.[2] A key policy question, therefore, is how to sustain or improve on this multiplier of ten, even as new technological innovations are ushering a major shift from a vertical model to a horizontal model of production and distribution in the communications and entertainment industries.[3] This task of adapting policies and regulations regarding the communications industry to new realities is made more challenging given its long legacy—one that goes back past Alexander Graham Bell to Benjamin Franklin, the first postmaster of the United States.

This introductory essay highlights selected issues discussed in the course of STEP's conference on Telecommunications and the New Economy.[4] The conference emphasized two transformations in communications: First, it emphasized the potential and challenges in the diffusion of broadband and Voice over Internet Protocol (VoIP). Second, it emphasized the transformation from vertical industrial organization in print, radio, entertainment, and broadcasting to more horizontal Internet based platforms. Speakers at the conference included industry-representatives, lawyers, and technologists, as well as some academics. They presented a variety of views on the challenges, opportunities, and policy prescriptions needed to sustain U.S. leadership in telecommunications.[5]

In this introductory summary, we first review progress in the measurement of communications equipment in the national accounts. We then look ahead to some emerging information and communications technologies and their possible contribution to sustaining the productivity improvements associated with the New

---

[2]See comments by Dale Jorgenson in the Proceedings section of this volume.

[3]Dale Jorgenson, "Concluding Remarks," in the Proceedings section of this volume.

[4]The enormous breadth of issues taken up at the conference leads to a tradeoff in the depth to which the conference or this introduction can cover them. We acknowledge this reality.

[5]At the same time, the conference was necessarily limited in time and focus. There are of course a variety of issues concerning the telecomm sector, not all of which can be addressed at any one-day event. For example, the conference did not cover a discussion of recent commercial history of the industry such as the dot-com boom and bust, the WorldCom fraud trials, and the legislative and legal history surrounding the 1996 Telecommunications Act. It also did not fully address all aspects of the impact of new forms of communications and media on regional economies and selected media markets. Another limitation was the relative focus on household use of the Internet and new media over business use of broadband, even though important productivity gains and economic advance often follow from business use of new information and communication technologies.

Economy. This then leads us to examine the reasons for the broadband gap in the United States and some alternative ways of bridging this gap. Finally, we highlight some of the policy challenges that emerge with "end of stovepiping" as information technologies and communications networks converge.

MEASURING TELECOM PRICES

How do new information and communications technologies translate into prices and hence consumer welfare? Mark Doms of the Federal Reserve Bank of San Francisco provided the participants of the STEP conference an overview of what the current official numbers say, and the challenges of coming up with good price indexes for communications equipment and services. He noted that while investment in communications in the United States had been substantial— around $100 billion per year, representing a little over 10 percent of total equipment investment in the U.S. economy—it had also been highly volatile. During the recession of the early 2000's, he noted, IT investment fell about 35 percent from peak to trough. (See Figure 1.) Doms noted that this recession might well be remembered as the high-tech recession, adding that "certainly what happened to communications played a major role in what happened to the high-tech sector."[6]

Measuring the dollars spent on communications technologies in the United States every year is difficult because the technology itself is rapidly changing. As demonstrated earlier in his study, a computer costing a thousand dollars today is a lot more powerful and versatile than a similarly priced one of 10 years ago— and this is no less true for communications equipment.[7] Twenty-five years ago, most long distance communications was handled through landline phones, in stark contrast to the diversity of means of communications in use today. As Doms' analysis points out, between 1996 and 2001 alone, there were tremendous advances in the amount of information that could travel down a strand of glass

---

[6]The rise of the Internet persuaded many investors in the late 1990s that demand for data-network backbone capacity was about to explode. Many anticipated Internet traffic to double every 100 days— a belief reinforced by an April 1998 report, "The Emerging Digital Economy," by the Department of Commerce, U.S. Department of Commerce, *The Emerging Digital Economy*, Washington, D.C.: U.S. Department of Commerce, 1998. Resulting large investments led to a fivefold increase in the amount of fiber in the ground. At the same time, technological advances increased the transmission capacity of each strand of fiber 100-fold, so total transmission capacity increased 500-fold. But over the same period demand for transmission capacity merely quadrupled, a rise that could easily be accommodated by existing networks. When it became clear that the predicted explosion of demand was not going to happen, operators frantically cut their prices, hoping to fill their empty pipes. Equipment-makers' sales collapsed and their share prices tumbled—leading to the burst of the telecom bubble. See *The Economist*, "Beyond the Bubble," October 9, 2003.

[7]Jack E. Triplett, "Performance Measures for Computers" in National Research Council, *Deconstructing the Computer: Report of a Workshop,* Dale W. Jorgenson and Charles W. Wessner, eds., Washington, D.C.: The National Academies Press, 2005.

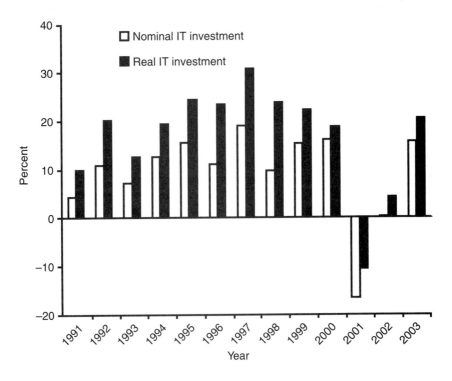

FIGURE 1 Annual percent change in IT investment.
NOTE: Percent changes based on year-end values.
SOURCE: Mark Doms, "The Boom and Bust in Information Technology Investment,"
FRBSF Economic Review, 2004: 19–34. Bureau of Economic Analysis.

fiber, adding that the price of gear used to transmit information over fiber fell, on average, by 14.9 percent a year over this five-year period. The fast speed of technological change makes the job of tracking prices complex because the capabilities of the equipment change dramatically under the same rubric of "computer" or "router." Whereas money spent on telecommunications was relatively easier to track 25 years ago when most purchases were of telephone switches, today's telecommunications equipment includes a wide array of technologies related to data, computer networking, and fiber optics.

Current methodologies for making inter-temporal comparisons in price and quality understate true price declines because they do not fully track these technological changes. While the Bureau for Economic Analysis has estimated that prices for communications gear fell an average of 3.2 percent per year between 1994 and 2000—in sharp contrast to the 19.3 percent fall in computer prices—Dom's analysis, which takes fuller account of technological changes, suggests that that communications equipment prices actually fell on the order of 8 to 10 percent over that period.[8]

While this new estimate is a step in the right direction, more refinement is necessary in measuring prices. As Doms notes that the job of keeping track of these developments is growing more difficult for statistical agencies, especially in light of their limited budgets and the rapid development of technology. "Unless the statistical agencies get increased funding, in the future, they are not going to be able to follow new, evolving trends very well," he concluded.

Indeed, as we see below, current trends in information and communications technology—benefiting from Moore's Law—will continue to disrupt incumbent businesses and traditional business models.

COMMUNICATIONS TECHNOLOGY: A VISION OF THE FUTURE

Moore's Law, which in its modern interpretation anticipates the doubling of the number of transistors on a chip every 18 months, has spurred the modern revolution in digital technologies for over forty years.[9] It is likely to continue for another ten to twenty years, according to experts in the semiconductor industry.[10] This pace of ever faster and cheaper semiconductors and semiconductor related technologies is likely to continue to have significant impacts, not least on communications technologies. As William Raduchel noted at the conference on telecommunications and the New Economy, the endurance of Moore's Law means that "the most powerful personal computer that's on your desk today is going to be in your cell phone in twenty years." Technologies for display, storage, and transmission of data are also expected to show rapid improvement, he added, though their rates of improvement are likely to abate sooner than that of semiconductors.[11]

---

[8]Mark E. Doms, "Communications Equipment: What has happened to Prices?" FRBSF Working Paper 2003-15.

[9]While by no means dictating an actual law, Gordon Moore correctly foresaw in 1965 the rapid doubling of the feature density of a chip, now interpreted as approximately every 18 months. Observing that the number of transistors per square inch on integrated circuits had doubled every year since the integrated circuit was invented, Gordon Moore predicted in 1965 that this trend would continue for the near future. See Gordon E. Moore, "Cramming More Components onto Integrated Circuits," *Electronics*, 38(8), 1965. The current definition of Moore's Law, which has been acknowledged by Dr. Moore, holds that the data density of a chip will double approximately every 18 months. Many experts expect Moore's Law to hold for another 15 years.

[10]See, for example, Robert Doering, "Physical Limits of Silicon CMOS Semiconductor Roadmap Predictions" in National Research Council, *Productivity and Cyclicality in Semiconductors: Trends, Implications, and Questions,* Dale W. Jorgenson and Charles W. Wessner, eds., Washington, D.C.: The National Academies Press, 2004.

[11]For a discussion by representative from these industries of the rate of technological change in these and other computer related industries, see National Research Council, *Deconstructing the Computer: Report of a Workshop, op. cit.*

Raduchel predicted that enhanced digital sampling, skyrocketing storage capacity, and expanded packet switching technologies will change the way we will work, communicate, and entertain ourselves in the future.[12] Faster computers mean that digital sampling for recording, playback, looping and editing of music will improve to the point where it is nearly error free, changing the way music is heard and distributed. Advances in storage capacity and speed will lead to new products (as already previewed with today's iPods and TiVos) that will likely challenge existing business models of how music and video entertainment is packaged and distributed, and ultimately consumed. In addition, advances in packet switching, where information is commoditized for transmission, will likely mean that "radio, television, classified information, piracy, maps, . . . anything" can be moved around a communications infrastructure with no distinction as to what they are. These developments, in turn, will require greater attention to the issue of standards that can allow for coherence as well as future growth and innovation.

These advances in capturing and distributing information and entertainment in commoditized packets build on the concept of the *stupid network*—where the intelligence is taken out of the middle of a communications network and put at the ends—a design principle that has already guided the development of the Internet.[13] According to David Isenberg, such an end-to-end network allows for diversity in the of means of transmission—including varieties of wired and wireless technologies—with this diversity creating greater robustness against the failure of any one element. As we see next, enhancements in packet switching capabilities are already making such novel technologies as Voice over Internet Protocols and Grid Computing technically and commercially feasible for widespread use.[14]

VoIP (Voice over Internet Protocol)

In Internet telephony, voice is broken into digital packets by a computer and conveyed over the digital network to be reassembled at the other end. The voice network of the future will run over the Internet Protocol, according to Jeff Jaffe of Lucent Technologies. Since this technology has a completely different capability than traditional landlines when it comes to voice quality, cost, and reliability, he predicted that it will bring about a generational change in voice communications.

Louis Mamakos of Vonage (a company that has introduced VoIP to commercial markets in the United States and elsewhere) cited two sources of opportunity that arise with VoIP: One is through sharing infrastructure, which comes from chopping up audio into packets and transmitting it over an existing packet-based

---

[12]See remarks by William Raduchel in the Proceedings section of this volume.

[13]David Isenberg, "Rise of the Stupid Network," *Computer Telephony,* (August):16-26, 1997.

[14]*The Wall Street Journal*, "Vonage plans to file for IPO," August 25, 2005.

Box A: VoIP—A Disruptive Technology

VoIP has the potential to undermine the business model underpinning the telecommunications industry. Factors such as the length of the call or the distance between callers, key determinants of cost today, are irrelevant with VoIP. In addition, VoIP augers more widespread use of videoconferencing as well as new applications such as unified messaging and television over Internet Protocol (IPTV).

Many analysts believe that the question is not whether VoIP will displace traditional telephony, but how quickly. This disruptive potential of VoIP is a challenge for telephone, mobile, and cable incumbents—with some attempting to block the new technology and others moving to embrace it.[a]

---

[a]*The Economist*, "How the internet killed the phone business," September 15, 2005. See also Dale W. Jorgenson, "Information Technology and the World Economy," Leon Kozminsky Academy Distinguished Lecture, May 14, 2004.

network, which yields significant cost advantages compared with traditional telephony. But equally powerfully, he contended, are opportunities that come from using software to provide a variety of services for the consumer. For example, by marrying it with the computer, phones could be programmed to control who can call through and when.[15]

Grid Computing

Grid computing, which allows users to share of data, software, and computing power over fiber optic networks is expected to be another major development in information and communications technology. Mike Nelson of IBM likens grid computing to a utility supplying electricity, noting that logging onto the Grid could provide a user access to far more computing power than is possible from a single computer system.

---

[15]"For the incumbent telecoms operators, though, what is scary about Vonage is not the company itself but the disruptiveness of its model. Vonage is a telecoms company with the agility of a dotcom. Everyone in the telecoms industry has heard of it, and has wondered what will happen if the model is widely adopted." See *The Economist*, "Between a Rock and a Hard Place," October 9, 2003. We many not have to wait much longer to see what will happen. See *The Financial Times*, "The internet's next big talking point: why VoIP telephony is quickly coming of age," September 9, 2005, which reports on the entry of Microsoft and Google into the VoIP market.

A widely known (but limited) instance of concept of grid computing is the current SETI (Search for Extraterrestrial Intelligence) @Home project, in which PC users worldwide donate unused processor cycles to help the search for signs of extraterrestrial life by analyzing signals coming from outer space. The project relies on individual users to volunteer to allow the SETI project to harness the unused processing power of the user's computer. About 500,000 people have downloaded this program, generating an amount of computing power that would have cost $100 million to purchase.

Grid computing is likely to have fewer nodes that are tied together than in the SETI case, said IBM's Nelson, but because the size of the machines can be larger—including large servers, storage systems, and even supercomputers—high levels of computing power can be generated. Further, since the systems involved in grid computing will be more tightly coupled and more general purpose, they can be far more versatile. The next step in grid computing, he predicted, is the "Holy Grid" where everything is connected to everything, running common software, able to tackle a wide range of problems. With the advent of such a grid, both small and large companies would be able to buy the computing power they need and get the software they need over this grid of network systems as needed on a pay-as-you-go basis.

In IBM's view, a part of the larger vision of Grid computing includes *autonomic computing*, where integrated computer systems are not only self-protecting, self-optimizing, self-configuring, and self-healing, but also come close to being self-managing. Another important component of this vision is *pervasive computing*, where sensors embedded in a variety of devices and products would gather data for analysis. These sensors will be located all around the world and the data they generate will have to be managed through the Grid. As Nelson predicts, "Soon we will have trillions of sensors, and that is what we really rely on the 'Net for."

The predicted arrival of Grid computing means that firms in the computer industry have an enormous stake in the future of telecommunications networks. With the Grid, the future of computing lies in complex network-based technologies, such as web services, which tie together programs running on different computers across the Internet, and utility computing to provide computing power on demand. With telecommunications firms becoming more dependent on information technology, and vice versa, the two industries are likely to become ever more closely intertwined.

Getting to the Future

While these and other emerging technologies offer alluring prospects for a more vibrant and productive future, a major focus of the STEP conference on telecommunication technologies concerned the regulations that condition the speed at which these technologies and others can be adopted as they become available. As Dr. Jorgenson pointed out in his introductory remarks, the issue of regulation is particularly germane to telecom, which is regulated at both the federal and state levels. Broadband regulation, in particular, was identified by several conference participants as a bottleneck to realizing the benefits of new information and communications technologies in the new "wired" and "wireless" economy.

SUSTAINING THE NEW ECONOMY:
THE BROADBAND CHALLENGE

Broadband, which refers in general to high-speed Internet connectivity, already supports a wide range of applications ranging from email and instant messaging to basic Web browsing and small file transfer, according to Mark Wegleitner of Verizon.[16] In the near future, he said, improved broadband networks can lead to true two-way video-conferencing and gaming as well as VoIP. The future of broadband, he predicted, includes multimedia Web browsing, distance learning and telemedicine. Beyond these applications, he noted, lay the possibility of immersive gaming and other types of information and entertainment delivery that comes with high band output combined with high-definition receivers.[17]

Can we indeed arrive at this promising future? Charles Ferguson of the Brookings Institution noted that while many foresee what a "radiant future" should look like, there exists an enormous gap for many between this vision for broadband-based technologies and the lack of adequate high-bandwidth access to a broadband network.

The Global Broadband Gap

Indeed, as many conference participants pointed out, the United States is falling behind other nations in access to high-bandwidth broadband.[18] Jaffe drew

---

[16]Individuals and businesses today variously connect to the nation's fiber-optic network through telephone lines (via digital subscriber lines or DSL), though television coaxial cables, and by fiber-to-the home, depending on the availability of these services within different jurisdictions. Wireless connections are also emerging as a viable alternative, as discussed later in the text.

[17]Many of these applications are already emerging, although the potential of many of these applications can be more completely realized through networks that are faster, carry more information, and reach more users.

[18]Commenting on a discussion of the United States slippage in broadband penetration rates, Dr. Kenneth Flamm of the University of Texas noted that it is important to carefully define what is meant by broadband. Broadband, he noted, describes a wide spectrum of bandwidth, with significant differences between its high and low end. In addition, he noted that while 99 percent of the U.S. population was connected by telephone or cable, and thus were potentially connected to the Internet, the issue of bandwidth size determined the types of applications that could be made practical to households and businesses.

Box B: The Demand Side of the Broadband Gap

With much of the discussion on how to address America's apparent lag in broadband adoption focusing on alternative models of service provision, the issue of broadband adoption among users has been relatively obscured. According to the Pew Internet Project's recent survey, the rate of growth in penetration of high-speed internet at home has slowed and could slow further.[a] While 53 percent of internet users had high-speed connections at home in May 2005, this level rose only modestly from 50 percent in December 2004. This is a small and not statistically significant increase, according to Pew's John Horrigan, particularly when compared with growth rates over a comparable time frame between November 2003 and May 2004 when the adoption rate rose from 35 percent to 42 percent. Dr. Horrigan concludes that there is less pent-up demand today for high-speed internet connections in the population of dial-up users and that this trend is likely to continue. He notes as well that currently 32 percent of the adult U.S. population does not use the internet at all, and that number is increasingly holding steady.

---

[a]John B. Horrigan, *Broadband Adoption at Home in the United States: Growing but Slowing*, Washington, D.C.: Pew Internet and American Life Project, September 24, 2005. Paper presented to the 33rd Annual Telecommunications Policy Research Conference.

attention to the reality that the United States had fallen far behind other leading nations in broadband penetration. Isenberg underscored this point, reporting that the International Telecommunications Union (ITU) had, in fact, ranked the United States in thirteenth place in 2003 and that the U.S. had likely since fallen to fifteenth place in broadband penetration. Citing the ITU figures for 2003, Ferguson reported that the penetration of digital subscriber lines (DSL) in the United States was 4.8 per 100 telephone lines, in contrast to South Korea where the penetration rate is 27.7 per 100 telephone lines. He noted that the United States had also fallen behind Japan and China in the absolute number of digital subscriber lines.

Acknowledging that this low figure for DSL is explained in part by the fact that a majority of U.S. residential broadband connections are through cable modems, Ferguson nevertheless contended that that this fact did little to change the overall picture. In the first place, he explained, when business connections were included, the percentage of total U.S. broadband connections provided by cable was relatively low. In the second place, even in the residential market the percentage of connections provided by cable had been holding roughly constant,

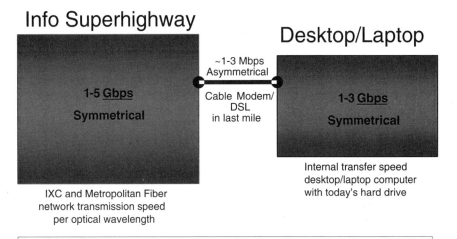

FIGURE 2 The broadband gap: Why aren't current services good enough?
SOURCE: Paul Green, FTTH Council Consultant.

as had the cable system's growth rate in respect not only to connections but also to bandwidth levels.

Ferguson observed that bandwidth constraints rather than computer hardware frequently dominate the total cost of adoption of a new network computing application. Personal computers were adequately powerful and relatively inexpensive, he noted, but given bandwidth constraints, deploying a high-performance, high-quality videoconferencing system or other applications could nonetheless prove extremely expensive.

Adding his own negative assessment of the U.S. competitive position, H. Brian Thompson of iTown Communications noted that while (what is commonly called) the Information Superhighway is capable of handling very high capacity in its fiber optic network, and while most desktops and laptops could function at between 1 and 3 gigabits per second, the problem was that there was often less than 1 megabit of connectivity between the two. This weak link—the broadband gap—was illustrated schematically by Thompson at the conference. (See Figure 2.)

In his remarks at the conference, Mark LaJoie of Time-Warner Cable cautioned that national aggregations showing the United States in thirteenth place worldwide do not tell the whole story. Differences in regulatory climate, the history and condition of infrastructures, the way in which products are used, as well as population densities are all factors influencing measures of broadband penetration. High-density cities like Tokyo and Seoul were likely to have higher levels of penetration, as do similar urban areas in the United States, he said, and added that while the infrastructure in Europe and Asia were newer, U.S. cable and telecom firms were making significant investments in expanding broadband capacity.

Agreeing that there are many ways to spin the numbers on broadband deployment, Mark Wegleitner of Verizon nonetheless acknowledged that "we aren't leading in what we have to perceive as one of the key technologies for any national economic environment going forward." He noted that his company, Verizon, was spending $12 billion annually on improving the broadband infrastructure—including expanding fiber to the home—thereby helping the United States catch up with other leading nations. At the same time, he predicted that "bandwidth demands are just going to grow and grow and grow," as new applications come into use.

Implications of the Broadband Gap

If broadband can serve as an engine for the nation's future growth and competitiveness, as emphasized by several participants at the conference, a lack of an adequate access to the broadband network may lead to a loss of this economic opportunity.[19] Assessing the impact of the broadband gap, Charles Ferguson noted that the "local bandwidth bottleneck" is having a substantial negative effect on the growth of the computer industry and of various other portions of the information technology hardware and software sectors. While conceding that computing an estimate of this impact in a rigorous way would be extremely difficult, he nevertheless asserted that "you can convince yourself easily that this effect is something on the order of one-half of 1 percent—or even up to 1 percent—per year in lost productivity growth and GNP."

Commenting on the national security implications of the broadband gap, Jeff Jaffe reminded the audience that the 9/11 Commission had recommended that the nation's digital infrastructure be prepared to deal with simultaneous physical and cyber attacks. In the case of a national emergency it will be important for first responders and other individuals to communicate effectively with each other and a high bandwidth, interoperable system is essential for this task, he said, adding that such a network is still not in place today.

---

[19]Dr. Raduchel, for example, noted that new technologies, like embedded sensors which rely on a capable broadband network, could emerge as the source of the next round of productivity improvements.

SOME EXPLANATIONS FOR THE BROADBAND GAP

While many of the participants at the conference concurred that the United States faces a broadband gap, views varied as to the reasons and well as solutions to this situation. Some suggested that the broadband gap has emerged because some telecom and cable companies have been reluctant to provide adequate interface between user and the fiber optic cable networks. Others suggested that the broadband gap arose from the consequences of federal and state regulations.

Flawed Market Motives of Telecom and Cable Companies

What is holding back high-bandwidth broadband penetration in the United States? Dr. Isenberg noted that the rise of the stupid network makes it difficult for the telephone or fiber company to sell anything other than commodity connectivity. In the new inter-networked model, it was the Internet Protocol's job to make all that was specific to a single network disappear and to permit only those things common to all networks come to the surface. Since the Internet ignores whatever is specific about a single network, including features that had formed the basis of competition for the telephone or cable companies, these companies have little to sell beyond access, he argued, and therefore faced little incentive in providing the public access to high-bandwidth broadband. The result, he said, was a crippled network with far less bandwidth available than technology would allow or than is available in other technologically advanced countries.

Ferguson suggested that flawed markets were behind the high cost of securing adequate bandwidth in the United States. He noted that both the telephone and cable companies had "severe conflicts of interests," and that they largely avoided competing with each other. Even competition for residential markets was "quite restrained, and much less substantial than you might suspect."

The conflict of interest for the telephone companies is "fairly obvious," Ferguson asserted. Incumbent businesses were providing very expensive voice and traditional data services. Very rapid improvements in price/performance of bandwidth would undercut their dominant businesses in a major way. The same was true of the cable system: It provided video services that could easily be provided over a sufficiently high-performance Internet Protocol network.

Consequences of Unbundling Network Elements

In the discussion following the second panel, Kenneth Flamm noted that more than one speaker had spoken of a tendency to dismantle some of the opening up of the local loop that had been the centerpiece of the 1996 Telecommunications Reform Act. The Act required incumbents to make parts of its network available to competing operators, in particular the "local loops"—the wires that run from telephone exchanges into homes and offices.[20]

The 1996 Act sought to promote competition by asking incumbents to share this part of their networks with rivals—technically known as "local loop unbundling" (LLU)—given that the expense for competitors to build their own networks would be very high in the short term. In practice, however, most incumbent operators saw unbundling as robbery, according to Thompson. This meant (as *The Economist* describes it) that "the incumbent must, in effect, give its rivals a hand as they try to steal its business. Not surprisingly, most incumbents find procedural, legal, and technical reasons for being slow about it."[21] Though intended to promote competition in the short run, local loop unbundling may have inhibited investments in alternate infrastructure that competitors might otherwise have made over the longer term. And because it forced incumbents to share their networks with rivals, this may have also deterred them from investing in new equipment. An unintended consequence of the 1996 Telecommunications Act may well have been to inhibit investment needed to provide high bandwidth broadband access over the local loop, although the issue of whether mandatory unbundling increases or decreases the roll out of broadband network access remains an open empirical question.

Even so, one of the authors of the Telecommunications Act of 1996, Charles Thompson, conceded that the concept of unbundled network elements, introduced in that legislation was moribund—that he "would be the first to put flowers on the grave of unbundled network elements."

Outdated Standards and Regulatory Uncertainty

Outdated standards and a regulatory uncertainty may be retarding progress in addressing the broadband gap, according to some conference presenters. On the issue of standards, Peter Tenhula of the Federal Communications Commission (FCC) acknowledged that wireless technology regulation was still being governed by a ninety-year-old spectrum management regime rather than one "rooted in modern-day technologies and markets." Such outdated regulations, he noted, fail to capitalize on technological advances in digital technologies such as those that allow for greater throughput of information, interference management, and spectrum sharing.

---

[20]Local loops can be either "legacy" copper loops or newer fiber broadband connections. The 1996 Telecom Act created considerable uncertainty for the unbundling broadband services. See, for example, the press release of April 8, 2002 by the Telecommunications Industry Association, "TIA Tells FCC That Unbundling Rules Discourage Broadband Investment," which recommends that the FCC not apply its network unbundling rules to new facilities used for the provision of broadband and high-speed Internet access services, and to apply them to legacy systems including copper loops, so as not to inhibit investment in wire-line broadband networks.

[21]*The Economist*, "Untangling the local loop," October 9, 2003.

Regulatory uncertainty is also holding down the installation of fiber all the way to the curb, noted Dr. Jaffe. Clear regulation is needed, he stated, to encourage sufficient near-term investment in fiber infrastructure. This regulatory environment may have been further clouded in recent years by increasing federal concerns about infrastructure protection, disaster recovery, and emergency services in the wake of recent concerns about terrorism. According to Jaffe, vendors such as Lucent face uncertainties in developing new products at a time when regulatory imperatives are very slow to come out.

Another important source of regulatory uncertainty is the patchwork of local regulation issued by individual municipalities. Cable infrastructure is often governed by city-specific franchise agreements, while telephone companies and other broadband providers may in some cases prefer statewide or even national authority as a means towards greater regulatory simplicity and predictability.

In addition, as Verizon's Wegleitner observed, prevailing uncertainties in updating regulation make it difficult for his company to invest in the development of an effective broadband network. Incremental rulemaking in the transition from the old regulatory regime to a new one often creates ambiguities, with investments of millions or even tens of millions of dollars hinging on the interpretation of words that, while written only a few years before, were already technically obsolete. "It is that interpretation that is going to determine the path forward of the network's evolution." This "unnecessarily complex regulatory environment," did not make sense in that it discouraged investment.

Thompson objected, however, arguing that large telecom and cable companies are not passive recipients of federal and state regulation and that, moreover, the current regulatory environment are greatly affected over the years by the power of incumbents on all sides. To the extent that incumbents influence regulation, the current uncertainty in regulation may well reflect the uncertainties that major cable and telecom providers are facing in coming up with a viable business model that allows profits in an arena that has been transformed by new technologies. Lisa Hook, recently of AOL-Broadband, noted in this respect that firms in the broadband industry were struggling at the service layer to find business models and revenue streams based on new technologies that would justify the investment needed to make nearly unlimited bandwidth widely available.

SOME ALTERNATIVE SOLUTIONS TO CLOSE
THE BROADBAND GAP

According to IBM's Michael Nelson, the Internet revolution is less than eight percent complete, with many new applications still to be enabled by future technologies like the Grid. Realizing this vision of the next-generation Internet will require both new technologies as well as significant investment, he cautioned, as it will entail providing whole neighborhoods with gigabits-per-second networks that are affordable and reliable as they are ubiquitous. "Getting there is going to require more intelligent, more consistent policies than we have today," he declared. Participants at the conference considered a variety of means by which the nation could close the broadband gap, of which some key approaches are previewed below.

Directed Government Incentives

Ferguson suggested that the nations that were ahead of the United States in broadband penetration shared two characteristics. The first was that their governments are "much more heavily involved in providing incentives and/or money and/or direct construction of networks than is the case in the United States." The second was that their Internet providers are under government pressure to improve their price and performance. For example, he said that the Chinese government had made it clear to the country's principal telecommunications providers that broadband deployment was a major national priority. The situation was similar in Japan and Korea, adding that government encouragement in Canada and the Scandinavian countries had also enabled those countries to surge ahead of the United States in high-bandwidth broadband penetration.[22]

For the United States, Ferguson recommended a variety of policy measures to bridge the broadband gap. Initiatives could include subsidizing the deployment of municipal networks and offering investment incentives to public and private providers. Putting more pressure on incumbents to open up their networks so that there is an open architecture broadband system that is more analogous to the structure of the Internet is another avenue.

Faith in Efficient Markets

In contrast to this more policy-driven approach, Verizon's Wegleitner noted that broader technical, financial, and regulatory improvements would reduce uncertainty and allow markets to function efficiently. While admitting that current challenges resisted simple solutions, he put forward what he called a short answer to the problem: "Let the markets rule." By this, he envisioned the Internet of the future as an *interconnection of commercial networks* such as Verizon's rather than the *confederation of commercial providers* that it is now. He added that the future requirements for services offered customers via broadband would be of such quality and scope that only an interconnection of commercial networks could provide this service.[23] To make this network of the future possible,

---

[22]For an assessment of Japanese policies to catch up and surpass the United States in Broadband connectivity, see Thomas Bleha, "Down to the Wire," *Foreign Affairs*, 84(3), 2005.

[23]The current Internet is based on a confederation made up of multiple service providers. Their ability (or inability) to maintain their interconnection arises from commercial issues, and not from the current design of the Internet.

Wegleitner recommended further development of appropriate standards for communication protocols and a new way of levying tolls on customers for use of the infrastructure that belongs to companies like Verizon, combined with a light regulatory touch.

Networks in the Hands of Customers

In the discussion that followed the first panel, Jay Hellman, a real estate developer, observed there exist business opportunities both in laying fiber to the home and making sure it functions. He likened the duo of fiber and services to a public roadway where service companies like FedEx and UPS competitively ply their fleets. It was desirable, he added. that the street be accessible to as many competitors as possible. He also added that his own frustration with the capacity offered by existing providers had prompted him to start his own small telecommunications company. Responding to this comment, David Isenberg noted that the development of technologies that allow customers to create their own networks and that create opportunities for individuals to provide service innovations was important to sustain innovation and provided a broader, more generic solution to the broadband challenge.

Municipally Owned Fiber

Thompson proposed a different approach, recommending the development of non-profit public-private partnerships at the local level to stimulate the development of broadband to the home. These partnerships would serve as a utility, lighting fiber but not provide any service on that fiber except those municipal services that the town or community base chose to provide. The network would be open to any and all service providers with an Internet Protocol basis—be they telephone companies, cable companies, software companies, or others providing online entertainment—and it would be used by all under the same terms and prices. Communities could build this network, just as municipalities build and maintain roads and sewers, he added. Citing the case of Ireland where, Thompson said, such partnerships have been successfully developed to provide broadband access.

While separating the network access component from retail services may help municipal providers of network infrastructure, more needs to be learned about the feasibility of this idea in the United States, including whether customers want to buy their services in this way. The issue of whether the municipal provision of infrastructure will in fact lead to more competition for broadband access also remains to be studied.

The Wireless Wildcard—A Silver Bullet?

Wireless broadband access can be a third tier that competes with cable and DSL, according to David Lippke of HighSpeed America.[24] In this way, wireless broadband can help overcome the limitations associated with traditional wired broadband access. While wireless broadband has been in limited use so far due to relatively high subscriber costs and technological limitations such as problems with obstacle penetration, rapid advances in technology are likely to overcome such challenges. Moore's Law applies to wireless no less than other forms of telecommunications, he noted, predicting that wireless data rates would reach all the points through which traditional telecom had passed.

In particular, scientists and engineers working on the upcoming WiMAX standard have resolved a number of problems that had bedeviled existing wireless protocols such as WiFi. The prospect of reaching gigabit speeds was now being mentioned, and other quality of service issues as well as lower costs of installation are being addressed. To the extent that these predictions are realized, the WiMAX protocol may well offer an effective wireless solution to the broadband gap, especially for smaller towns and communities across the United States.

THE END OF STOVEPIPING

The move from analog to digital information and communication technologies is ushering a major transformation disrupting how telecom, cable, and music and video entertainment companies, among others, do business. Because analog solutions were all that existed until recently (except in some fields of computing), these industries each matured into separate industries, with separately evolved business models and regulatory frameworks. In the digital age, however, basic technologies like digital sampling and packet switching enable the commoditization of voice, data, and images into digital packets that resemble each other. These packets can be sent over the Internet with no distinction as to what they are, to be reassembled at the intelligent ends of the network.

Drawing on these observations, William Raduchel noted at the conference that the information and communications technology revolution will usher the end to stovepiping as service and content providers shift from vertical integration to a greater reliance of horizontal platforms. This change, he noted, will give rise to a variety of public policy issues as individuals and businesses in the economy restructure to take advantage of the potential offered by new technologies.[25] He

---

[24]Also mentioned at the conference was broadband over power lines, which at the time was being reviewed by the FCC.

[25]A key example of contemporary relevance is the offshore outsourcing issue. For a discussion of this issue, see National Research Council, *Software, Growth, and the Future of the U.S. Economy: Report of a Workshop*, Dale W. Jorgenson and Charles W. Wessner, eds., Washington, D.C.: The National Academies Press, 2006. See also Catherine L. Mann, *High Technology and the Globalization of America*, forthcoming.

Box C: Some Factors Affecting the End of Stovepiping

While the digital transformation has the potential to disrupt traditional vertically-integrated industrial organizations, some factors may inhibit a transformation to a fully horizontal platform.

- **Open Network Architecture**: The horizontal organization of communications requires a relatively open network architecture. However, if systems or content providers do not have access to physical or logical pipes, those providers cannot reach their customers.[a]
- **Separation of Carriage from Content**: Some customers may prefer to purchase services in bundles that include access, as noted by Lisa Hook. Here, vertically-integrated firms may have a competitive advantage over firms that supply pipes or content exclusively.
- **Social Policies that Favor Universal Access**: Where social policies set access price below a competitive market price, the supplier of the access must also be able to cover its total cost from the supply of some other higher-margin services or receive a subsidy.
- **Economies of Scope**: There may be economies of scope between providing communications services and network facilities.

---

[a]Consider, for example, the FCC's Video Dialtone initiative in the 1990s, which attracted substantial investment from incumbent telephone companies until it was determined that some portion of the bandwidth had to be made available to competing content providers. For a wider discussion of the limitations of open access cable, see Thomas W. Hazlett and George Bittlingmayer, "The Political Economy of Cable 'Open Access,'" *Stanford Technology Law Review*, 4, 2003.

also noted that the speed of change is likely to be such that the economy may not be able to adjust to it readily. Among the issues to be addressed is the challenge to intellectual property rights and question of regulation, which is expected to be very challenging.

The potential and implications of the move from analog to digital information and communication technologies were discussed by several of the conference's participants. Key points from these discussions are summarized below. As in any conference that includes a variety of perspectives, some of these policy recommendations are mutually contradictory, and evidence may be required regarding their efficacy.

Convergence and Competition

Raduchel sees the Internet as having two complementary aspects—it is both a physical set of networks as well as a protocol known as TCP/IP. At the present, the physical network can only support movies and other applications at low bit volumes and is often not cost effective—although this can be expected to change as technology improves and the broadband gap is overcome. The significance of the Internet Protocol, he said, is that it makes all networks look the same and allows interoperability. It was for this reason that the telecommunications world could be expected to move to one set of interconnected webs, he said, predicting that "5 to 10 years from now, we will be online all the time."

This convergence is challenging the traditional business models of firms in these industries. How would telecom companies, for example, deal with new technology that makes cell phones work perfectly everywhere or with much cheaper VoIP service? The next decade, warned Dr. Raduchel, would be marked by "lots of dislocation" as firms attempt to adjust to new technological and commercial realities.

According to Mr. LaJoie, the convergence of data, voice, video, wireless, public networks, and private networks in an end-to-end infrastructure was changing the terms of competition across industries. Where there was once a big separation between what the telecom and cable industries did for example, "now everybody is in everybody else's business." While cable television, Internet, Cellular, WiFi, and Satellite transmission businesses were once distinct, LaJoie believes that they are all destined to overlap and offer similar kinds of products, suggesting with some optimism that the economic rewards that will arise from this competition would be what drives continued innovation, the advent of new services, and increased broadband connectivity.

The potential end of stovepiping also poses new challenges for consumers. Many consumers, faced with a proliferation of Internet services, operating systems, and devices will want a service that is easy to use and integrated, predicted Ms. Hook. She noted that companies like AOL Broadband see a market opportunity as aggregators, packaging a variety of content and communications services over the Internet and protection against viruses and spy-ware that are easy to launch and use.

Intellectual Property in the Era of Digital Distribution

In addition to disruption in the business models of firms that deliver a digital signal is the disruption to business models of firms that provide the content. Indeed, the music and entertainment industries are among those that are also undergoing a fundamental shift in the digital age. Andrew Schuon of the International Music Feed television network noted that while the public's desire to consume music has never been greater, with new technologies allowing users to take an entire music collection with them anywhere they go, the key problem for content providers is how to make money selling music in the new medium— given that technology already available has allowed consumers to share music and other content with each other for free. At present, he noted, legitimate downloads account for only a few percent of all downloads from the Internet.

He noted that technology developed for building legitimate services makes it now possible to protect intellectual property, to monetize it, and to track licenses while, at the same time, creating a good experience for the consumer. However, this technology has to catch up with consumer expectations that have developed in the absence of such constraints: "If you steal the content, you can do anything you want with it—put it into any portable device, put it on as many computers as you have, use the content as you see fit." The challenge for the music industry is to find a way to get the consumer to pay for its product while at the same time being more creative than the illegitimate download sites. The music industry, Mr. Shuon said, has to offer the modern customer the flexibility to use the content in the way they want to, in addition to offering superior content and a fair price.

Steve Metalitz, of the law firm Smith and Metalitz, agreed that developing a legitimate market for copyrighted materials over broadband—for entertainment, services, software, video games, research, and reference works—was indispensable for the long-term viability of these industries. Acknowledging that piracy will continue to be a problem, he added that the challenge for the future of broadband is to achieve a relatively low level of piracy and a very high level of legitimate products. Addressing this challenge requires:

- Developing legitimate markets for copyrighted materials over broadband;
- Providing greater security for delivering content to an end-user including measures to ensure that the income-generating potential of material going into the pipe did not vanish forever;
- Creating a usable legal framework to protect the technological measures used to control access to copyrighted material in the network environment;
- Focusing enforcement of piracy problems on organized criminal groups as well as dedicated amateurs who play a role in making the system insecure; and
- Improving public education to make consumers aware that certain types of file sharing is illegal and the need to secure permission to avoid copyright infringements.

Cooperation, Mr. Metalitz concluded, is needed among providers of network services along with better communication with policymakers to advance these objectives.

The Challenge for Regulation

According to Peter Tenhula of the FCC, the challenge for regulation concerns the migration from decades of regulatory stovepipes towards a new vision

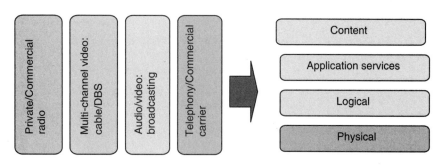

FIGURE 3 Vertical silos to horizontal layers.

of a variety of applications and services (covering voice, video, and data among others) that are provided over multiple and competing telecommunications platforms (including cable, satellite, DSL, and power lines). For this idea to work, content or service providers need a choice of mechanisms by which they can reach their customers. Rather than preserve the artificial vertical integration that had existed for decades and had created silos that grew up over the years, Mr. Tenhula suggested that it made better sense to let the natural layers fall as they might. (See Figure 3.) Replacing sector-specific communications regulation with a layered regulatory model, he added, would better complement the networked characteristic of the New Economy.

The FCC's agenda, he said, was to guide and propel the journey from a slow, conventional analog world to a digital world with significant opportunities for faster, more reliable, higher quality information and communications, with the overall goal of providing substantial benefits for American consumers.

CONCLUSION

Concluding the conference, Dale Jorgenson noted that the New Economy had witnessed a huge shift from a vertical model to a horizontal model in the computer, semiconductor, and communications industries. In this new model, he said, most of the interesting innovations were disruptive. The challenge for businesses in this changing environment was to figure out how to make money, which was hard given that consumers were both clever and unpredictable. It was "too bad," he said that the consumer ends up carrying away most of the welfare, which then cannot be delivered to shareholders. But in another respect, he added, the fact that "consumers emerge over and over again as the big winners . . . [is] a great thing about the New Economy."

Jorgenson characterized the policy issues in the telecommunications challenge as particularly difficult. While many economists are prone to offer private property as an answer to policy dilemmas, the presence of common property in the form of the digital communications infrastructure made matters more complex, he noted, adding that a way had to be found of maintaining common facilities within a market based approach. The transmission of property such as data, software, and music across this network also raised questions about its protection, while ensuring privacy for users. Taken together, these issues provide a robust agenda for further study and consideration about the New Economy—which, he noted, has been a central aim of the National Academies' Board on Science, Technology, and Economic Policy.

II

PROCEEDINGS

Introduction

Dale W. Jorgenson
Harvard University

Dr. Jorgenson welcomed the audience to the fifth in a series of symposia sponsored by the National Research Council's Board on Science, Technology, and Economic Policy (STEP) and devoted to the theme "Measuring and Sustaining the New Economy." To counter the view that the New Economy had "disappeared in the year 2000," he pointed out that recent figures indicated that, since the end of the previous recession in 2001, productivity growth had been running about two-tenths of a percentage point higher than in any recovery of the post-World War II period. He therefore applied the label "alive and well" to the New Economy, which, he noted, STEP had been tracking almost from the time the phenomenon was recognized in the policy community.

Reviewing STEP's previous symposia on the New Economy, Dr. Jorgenson recalled that the first had taken place in the year 2000 and had resulted in the publication of *Measuring and Sustaining the New Economy*. The thesis of that book—which, he said, still held up very well—was that technology is the main source of the development denoted by the term "New Economy," and that the key technologies center on semiconductors. The second symposium, addressing semiconductors specifically, dealt with speed at which semiconductor technology develops, which is described by Moore's Law. That topic is critical to telecommunications technology, the focus of the current symposium, just as it is to the development of technologies related to computing.

The subject of the third symposium in the series, which was held about a year after that on semiconductors and for which a report was yet to be published, was computers. According to the future it sketched out for Moore's Law, the accelerated pace of development of semiconductor technology that started in the mid-1990s was to continue; and, in fact, the rate of progress between that meeting and the present had proved this expectation correct. Addressing those in the audience who tracked semiconductor technology, Dr. Jorgenson observed that Robert Doering of Texas Instruments and other leading authorities in the field had projected semiconductor development to continue at that accelerated pace for at least another decade or so, something of extreme relevance to the day's topic. The fourth symposium of the series, held in February 2004, examined developments in software technology, which he described as a "much less tractable topic." STEP will publish a report on that meeting as well.

THE NEW ECONOMY: A COMPREHENSIVE PICTURE

Taken together, the work sponsored by STEP under the rubric Measuring and Sustaining the New Economy—examining the specifics of semiconductor technology, the base technology driving the pace of technological development; computing; software; and, at the current meeting, telecommunications—has produced the most detailed and comprehensive picture available to date of what is known as the New Economy. And true to the components of STEP's name, this study had encompassed economics, about which a great deal had been learned in its course, as well as technology, which had been the focus of much of the deliberation.

Turning to policy, Dr. Jorgenson raised the issue of which policies condition the speed at which new technologies are adopted as they became available. This is particularly germane to telecommunications, which is regulated not only at the federal level, a subject to be addressed shortly by Peter Tenhula of the Federal Communications Commission, but also at the state level. And, in addition to the involvement of many different regulatory bodies, there is that of Congress, which has passed major legislation, and the courts, which have always been the arena of last resort. So STEP—the Board on Science, Technology, *and* Economic Policy—was bringing all three together in the day's symposium.

Thanking Dr. Charles Wessner of the STEP staff and Dr. Bill Raduchel, a member of the Board, for organizing the symposium, Dr. Jorgenson turned the floor over to the latter, the subject of whose talk was the end of stovepiping.

Overview: "The End of Stovepiping"

William J. Raduchel
Ruckus Network

Dr. Raduchel recalled that he had joined the STEP Board around the time of the first of the workshops on Measuring and Sustaining the New Economy and noted his pleasure at watching the Board's journey down the path that Dr. Jorgenson had described as it tried to increase its understanding of the forces shaping the economy.

To get his talk under way, he called attention to a recent agreement between Twentieth Century Fox and Vodaphone under which the studio would develop one-minute original episodes of the television show *24* for distribution beginning in 2005 over cell phone handsets in the U.K. He expressed doubt that, 5 years before, many people would have imagined themselves watching an episode of a television show that was not going to be available via television, let alone that there would be a major new form of entertainment having mobile handsets as its platform. He said he expected these one-minute episodes, which in and of themselves he found mind-boggling, would be available on Verizon Wireless in the United States later in 2005. With others working on all sorts of things to distribute via mobile phones, he said, it was a question as to what the real market would turn out to be. Citing the Fox-Vodaphone deal as an example of "the ultimate in how convergence is happening," he underlined the difficulty of predicting "what's going to drive the world going forward."

TECHNOLOGY: SIX MAJOR THEMES

Dr. Raduchel briefly introduced the six themes that he would address in his presentation:

- **Innovation happens when it can**, which means that if technology enables it, someone is going to do it—a fact that, he said, "people who study technology really do understand."
- **Digital sampling and packet switching are fundamental changes**, and a new wave that would bring the latest of their many impacts was just becoming visible.
- **Technology change is not over**, as Dr. Jorgenson's words had just evidenced.
- **All networks collapse to one set of interconnected webs**, as he had observed on a recent visit to South Korea. That country was on course to be entirely lit by 2006, so that the user would be able to go seamlessly from EVDO[1] to a specifically Korean WiMax standard that was evolving. "If you're Korean and you have the right gear," he remarked, "you will be online everywhere you go in the country."
- **Convergence is coming.**
- **Regulation is going to be very challenging.**

Innovation Happens When it Can

Among practical examples of what science has enabled have been telephony, records, radio, TV, mobile telephony, CDs, and DVDs. Students of information technology understand that specialized solutions are possible years or decades before generalized solutions, Dr. Raduchel observed, so that applications will always emerge "in a way that looks unique in the beginning but over time becomes blended in with the overall theme of information technology." Because analog solutions (which was all that existed until the 1980s except in some limited fields of computing) were available before digital solutions, each developed into a separate industry. In truth, however, they are not separate.

Digital Sampling

To evoke this phenomenon, Dr. Raduchel harked back to high school math class, when students learn to approximate a curve by placing rectangles underneath it. If the rectangles used are sufficiently narrow, the approximation can be so close that its difference from the actual curve is negligible. The principle in

---

[1]EVDO or Evolution Data Only or Evolution Data Optimized is a fast wireless broadband access (3G) without needing a WiFi hotspot. For additional information, access *<http://www.evdoinfo.com>*.

digital sampling is similar: Instead of trying to keep track of a curve of, say, sound or video, one approximates it. As computer speed increases, the approximation improves to the point that the reproduction attains the same quality, and—something very important from the viewpoint of the consumer—it tends to be error free.

Unanticipated Consequences of Change

Such advances can have unanticipated consequences that many have difficulty perceiving except in retrospect. Drawing an illustration from the recording industry, Dr. Raduchel recalled inviting friends over to hear the "virgin play" of a new vinyl record, which could be 20 to 40 percent better than any subsequent play, depending on the quality of the phonograph. "Music was a form of primary entertainment," he said, "because people would get together and listen, since that first play was so special." But the one-hundredth play of a CD is the same as the first, so it no longer matters whether one is hearing it the first time, or the hundredth time, or the twentieth time. He speculated that recorded music has lost its place as a primary source of entertainment because, through a change in technology, the special appeal of listening to it the first time has disappeared.

Packet Switching

This technology, the foundation of the Internet, applies the same basic idea. A signal is transmitted over the air or through a wire as small packets that are then reassembled at their destination. This process commoditizes information, since all forms of it are turned into packets and each packet resembles the next. All that is done by this huge worldwide network, the Internet, is to move the packets around without distinction as to what they are. "They can be radio, television, classified information, piracy, maps, anything," Dr. Raduchel stated, adding that "everything is just bits" in the world that has resulted from this "very profound technology change."

Technology Change is Not Over

Dr. Raduchel noted that Intel had recently made public its engineers' prediction that the minimum 30 percent annual rate of improvement sustained by semiconductor performance for the previous two decades would remain a constant for at least 10 and, possibly, 20 more years—that is, that Moore's Law would continue in force. The result of maintaining this rate of improvement, which equates to 97 percent per decade, is that "the most powerful personal computer that's on your desk today is going to be in your cell phone in 20 years." And recalling presentations at an earlier symposium in the current series, "Deconstructing the Computer," he said that display, storage, and transmission could be expected to show even more rapid improvement, although their rates of improvement were likely to abate sooner than that of semiconductors themselves. "In general," he stated, "we will probably see a two-order-of-magnitude drop using conventional technology in computing, transmission, and storage."

Innovations that were in the offing for the following 3 years would prove both interesting and disruptive. For ultrawideband wireless, familiar to technology watchers as USB/1394, standards had been agreed, and products would hit the market the following year. These technologies were described by Dr. Raduchel as "a way of going from your personal computer to your TV set seamlessly, wirelessly, instantly." Also coming was wireless broadband beyond WiMax, "one of the most fascinating developments" and among the topics to be addressed by another of the day's speakers, Dave Lippke of HighSpeed America. WiMax itself was capable of speeds up to 250 Mbit per second—"really high-speed transmission," he observed, "lighting the whole country."

Storage Capability Skyrocketing

The advances in storage would be as large as those in any other technology. The capacity of the serial ATA disk drive, representing the newest generation of that product, would grow to approach terabytes in size over the following 5 years. Because the price of a disk drive had frozen at around $80, producers competed through growth in speed and capacity. It was for this reason, Dr. Raduchel pointed out, that the music industry was so nervous. "The personal computer you buy in 3 years will be able to hold every song ever made," he predicted, "and still have a lot of room left on its disk drive." Meanwhile, hard drives would appear that were small enough to fit into cell phones but could store gigabytes of data.

Dr. Raduchel then turned to silicon tuners, which he said provide the ability to tune television signals off a satellite, over the air, or over cable. Noting that these devices become cheaper as they are moved into small computers, he said it would soon be possible to record 16 channels simultaneously. "Those of you with TiVos," he advised, "think 'TiVo on steroids.'"

With these changes, the cost per bit keeps dropping. An e-mail is, in general, a few thousand bytes; a song, about 4 megabytes; a DVD movie, about 5,000 megabytes; and an HD movie, about 50,000 megabytes. That, for instance, the HD movie is 10 times the bytes of a DVD movie—but is still a movie—indicates that the value per bit being transmitted has, with the move into entertainment, declined massively from where it was when the Internet started.

All Networks Collapse to One Set of Interconnected Webs

The Internet itself is two things, a physical set of networks and a protocol known as TCP/IP, both of which were designed mainly for e-mail. While the Internet was workable for movies at low volumes, it was "not yet cost-effective over the long haul"—although, Dr. Raduchel noted, "that could change." Broadcast remained very efficient as a means of delivering large volumes of bits and, combined with large numbers of hard drives, "they begin again to approximate the same thing. That's one of the battles you're going to see in the next 5 years." The strength of Internet protocols, however, is that they make all networks look the same and allow interoperability, and it was for this reason that the telecommunications world could be expected to move to one set of interconnected webs. "Five to 10 years from now," he predicted, "we will be online all the time."

Voice providers understood that their industry was on the verge of becoming a feature, something not without precedent in the technology sector. "If you're in the industry," Dr. Raduchel stated, becoming a feature "is not good." By way of illustration, he pointed to Skype, a company employing eight programmers in Estonia that had become a provider of international telecommunication services and had grown to the point that it was disrupting the industry in the United States. And on the way was a generation of phones that would allow users to roam to 802.11b or 802.11g networks, the wireless networking, or WiFi, that had become common in hotels, offices, and homes.

Days Numbered for Landlines?

This would not be as minor a change as it might seem, for it would improve cell phone service significantly in suburbs, where resistance to the placing of cell towers had been common. The many people who had been holding onto landlines because their cell phones did not work well in their homes would suddenly be able to roam to a broadband connection and have their cell phones work perfectly. He called this prospective development "a major threat to the established telecoms" because, as he said: "If your cell phone works perfectly, why do you use anything else?" And every major player had entered the market for another form of very cheap telephony, voice over IP, or VoIP. They were not sure how they were going to make money, but they were sure that they'd better be there. Speakers from both Verizon and Vonage were to address the subject later in the day.

Dr. Raduchel's current professional activity involves serving college students, who represent the next generation of technology users. "They live on their PC and their cell phone," he said, explaining that their primary music and video device is the former, and that their main communication takes place via instant messaging and cell phone. The students spend about 6 hours a day online as opposed to less than 6 hours a week watching television in the traditional sense; live sports account for half of that viewing time. They almost never pay for media. "They see everything as a victimless crime and don't worry about it," he observed, noting that "darknet copying abounds: inside the dorms, where you have very high speed network connections, these kids copy everything and copy it a lot."

Convergence is Coming

Dr. Raduchel stated that mobile phones would be "the first truly converged devices." It is expected that, as of 2006, one-third of cell phones in South Korea would be able to receive 13 video channels, 25 audio channels, and 3 data channels via direct-to-mobile broadcasting (DMB). Broadcasting would be from *s*-band satellites, employed in the United States by XM and Sirius, to cell phones in cars. Soccer was to be among the offerings on the video channels, but otherwise programming had not yet been set. In addition, SIM cards like those used in GSM phones were expected to be made available to cell phone users by South Korean banks; installing the card would equip a phone with a fingerprint reader linked to the bank, thereby turning it into a banking terminal. "The mobile phone will begin to become the dominant way of conducting transactions," he asserted, "because it will be more secure, more reliable, and easier to use than anything else out there."

Consumer broadband would follow the cell phone as a vehicle of convergence; and, in fact, this process had already gotten under way with voice over IP. In television, the newest competitors were Dell and Hewlett-Packard, which were accustomed to coping with much thinner margins than, and were able to produce in high volume better than, consumer electronics companies. "There is no difference between a flat-panel television and a PC except the packaging and the software," said Dr. Raduchel, "so Dell and HP represent major threats to these industries." He again pointed to the entry of Skype into competition for global long-distance services.

Public Policy Issues Straightforward

The public policy issues looming over the landscape of convergence, Dr. Raduchel said, were relatively straightforward:

- **The speed of change was such that the economy was unable to adjust to it readily.** "You can't have this much change in this little time without having lots of disruption," he opined.
- **Increased options for consumers were being traded off against the loss of capital and jobs.** The outstanding fixed debt of telecommunications firms, he said, had reached around $60 billion or $70 billion worldwide.
- **Intellectual property rights (IPR) were a widening concern.** While music and films were in the spotlight, the challenge to IPR had reached everything that could be copied.
- **Growing complexity had its cost to the economy.** Pain was a related issue, he said, pointing to a *Wall Street Journal* column in which Walt Mossberg explained why the PC is the consumer device on which we are most dependent and that we most hate.

• **Security and reliability, while they might seem far-fetched concerns, were very real.** That consumer PCs connected to broadband could be turned, by an attacker unleashing them all simultaneously, into a massive weapon against the U.S. economy was "a doomsday scenario [but] not an implausible doomsday scenario." Dr. Raduchel recalled that the counterterrorism expert Richard Clarke, while working for the U.S. government, had been "passionate" about the risk that a so-called distributed denial of service attack might pose. This prospect, which casts "Microsoft Windows as the greatest threat to national security that exists today because of the degree of vulnerability in it," was the source of great worry among experts, he said, adding: "I don't know what we can do about it."

Regulation Is Going to Be Very Challenging

The questions of how these industries-turned-features were to be regulated, and of who would do it, were very profound.

As was typical of the STEP symposia on Measuring and Sustaining the New Economy, Dr. Raduchel reflected, there was virtually no possibility of resolving all the issues aired, but there was an opportunity to do a good job of beginning to frame the questions that should be asked about them. Then, thanking the audience, he turned the podium back to Dr. Jorgenson.

Remarking that Dr. Raduchel's presentation had set the stage for a discussion of policy, Dr. Jorgenson proposed leaving comments and questions until after the following speaker, Peter Tenhula of the Federal Communications Commission.

Technological Change and Economic Opportunity: The View from the Federal Communications Commission

Peter A. Tenhula
Federal Communications Commission

Mr. Tenhula expressed his pleasure at having been invited by the STEP Board to present a perspective from the Federal Communications Commission (FCC) on the telecommunications challenge the symposium was exploring, although he also voiced regret that Michael Powell, the Commission's chairman, had not been able to attend and to share his views. He speculated that Mr. Powell, along with other FCC staff, were on their way to Nashville to attend the annual convention of the National Association of Regulatory Utility Commissioners, whose members are the state telecom regulators. As the Commission had ruled the previous week that the states could not regulate Internet telephone services and other IP-enabled services, the convention would "probably be like a lion's den" for those representing the FCC, he said. "But knowing the chairman as well as I do, I predict that he will be as safe as Daniel among those lions"—even if he would most likely have preferred attending STEP's symposium instead.

Having spent 5 years as Mr. Powell's senior legal adviser, which Mr. Tenhula called "by far the most valuable experience" to date of his 14 years at the FCC, he was now working on spectrum policy reform. Warning that he might spend a disproportionate amount of time on spectrum, he apologized in advance, while also promising to try to shed some light on the wireless issues raised by Dr. Raduchel. He said he would do his best to depict in full the attempt of the FCC, a "70-year-old regulatory agency . . . to catch up and keep up with technological changes."

In light of both the subtitle of his presentation, "The View from the FCC," and his position as a career employee, Mr. Tenhula said he wished to call particular attention to the agency's standard disclaimer: "This presentation and the views expressed by the presenter do not necessarily reflect the views of the FCC, the chairman, individual commissioners, the FCC staff or the administration."

Mr. Tenhula said he would begin by highlighting how the FCC was embracing and fostering technological change and innovation, borrowing a metaphor from Mr. Powell that had been guiding the Commission for the previous few years: The Great Digital Migration. He would then briefly describe how the FCC's policies were fostering economic opportunity and entrepreneurship in line with another of Mr. Powell's themes, Power to the People, and with the ideal of consumer-driven innovation. Then he would discuss wireless as a successful deregulatory model for telecom, along with some current challenges of spectrum policy reform. Finally, he would share his personal take on some of the future issues and opportunities being generated by new technologies and on the regulatory, technical, and economic challenges that the FCC, Congress, the administration, the telecom industry, and academia might be in a position to address.

THE GREAT DIGITAL MIGRATION

To illustrate the challenging regulatory transition dubbed by Chairman Powell as "The Great Digital Migration," Mr. Tenhula projected a graphic representing the "winding road from decades of regulatory stovepipes to a heavenly vision of a wide variety of applications and services being provided over a plethora of competing telecommunications platforms" (see Figure 1). Mr. Powell, he said, had been framing the FCC's agenda around this concept with the intention of properly guiding and propelling the journey from a slow, conventional, analog world to a digital world with significant opportunities for faster, more reliable, higher quality information and communications. Accompanying this change would be an inevitable and radical transformation resulting in substantial benefits for American consumers. While he acknowledged that there were potholes depicted along the road—something he associated with the quality of roads in Washington, D.C.—he stressed that the vision adopted by Mr. Powell was a "very optimistic" one. "He could easily have adopted the more pessimistic image of an impending train wreck," Mr. Tenhula pointed out.

Mr. Tenhula expressed one of this approach's two guiding principles through the phrase "multiple platforms are the key." The goal was to have additional physical platforms generate tremendous consumer benefits by delivering multiple facilities-based competitors. The transformative impact would scream out for deregulation because it would be accompanied by:

1. increased need for risky and heavy capital investment;

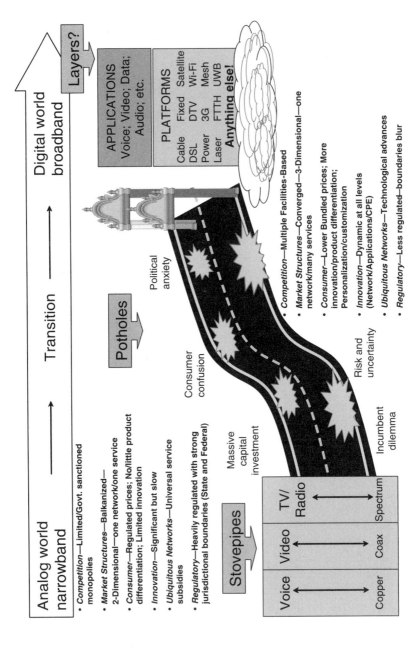

FIGURE 1 The great digital migration.

2. demonstrable positive competitive effects and better ways to deal with demonstrable competitive harms; and
3. the anxiety and uncertainty of never-ending regulatory proceedings.

The initial step under the second guiding principle, that of "avoiding reflexive symmetry," was to allow significant development, then assess market conditions and determine what, if any, regulation might be appropriate. A good example of this, wireless, will be discussed further in a moment. In the wake of the initial step, it would be important not to burden new services and applications with the physical-layer-based stovepiped regulations of yesterday.

Mr. Tenhula listed some **Wireline Migration** matters in which the FCC was involved:

• The **Triennial Review**, a follow-up to the 1996 Telecommunications Act, was a "very litigious" proceeding involving deregulation of local loops as well as delegation to and regulation by the states of unbundled network element and its pricing.

• The **Brand X Appeal**, in which a petition of *certiorari* was pending before the U.S. Supreme Court, involved the appeal of a Ninth Circuit case dealing with whether cable-modem service was defined as an "information service" or a "telecommunication" service under the Telecommunications Act.[2]

• A **Universal Service Review** was undertaken by the FCC in order to ensure the ubiquity of the next generation of technologies, as intended under the original 1934 Telecommunications Act.

• **Intercarrier Compensation** dealt with how carriers compensate each other for exchanging traffic, as well as how to deal with legacy access charges and other approaches to compensating carriers participating in the networks of networks.

• The **IP-Enabled Applications/Services** proceeding—covering all applications over Internet Protocol, notably voice over Internet Protocol (VoIP)—was then pending. The FCC, as Mr. Tenhula had mentioned, had indicated where its jurisdiction lay in a couple of cases, having told the states to stay out of VoIP and other IP-enabled applications in a ruling on the Vonage case just the week before. As yet unresolved, however, were many other, tangential issues, among them Universal Service, Communications Assistance for Law Enforcement (CALEA), and Enhanced 911 (E911).

• **Broadband over Powerlines**, a rule making recently taken up by the FCC, dealt with the provision of very high speed bandwidth over the electrical grid. The issue facing the Commission here, rather than whether to allow this new broadband competitor, was the leakage of signals from power lines that interfered with other spectrum-based services.

---

[2]The Supreme Court since ruled in the FCC's favor on June 27, 2005.

He then turned to examples of issues concerning **Mass Media Migration** that were occupying the FCC's attention:

• **Digital Television Transition** and **Cable Must Carry** were tied together. The FCC, which was contemplating a "hard date" for the conversion from analog to digital high-definition television, was looking at a time frame of around 2009, but the staff's proposal was "very tetchy." Also controversial was the main question regarding what cable must carry: How much of that digital signal must cable operators carry?

• **Direct Broadcast Satellite (DBS)** providers EchoStar and DirecTV had proved effective competitors against the cable video platform, while other high-definition, focused, direct-broadcast satellite competitors—VOOM, for one— were getting set to come online.

• **Digital Satellite Radio**, as represented by XM and Sirius, was "taking off like gangbusters."

• **Digital Terrestrial Radio**, also called "high-definition radio," would bring crisp, clean, CD-quality music to homes and automobiles.

• **Media Ownership** battles were continuing.

• **Video Over DSL** was showing a great deal of promise.

• **Wireless Mobile Media**, an example of which was the one-minute TV episode had been the object of recent announcements by Texas Instruments, Qualcomm, and Infineon, which also planned to transmit high-quality video to the small screens on cell phones.

Observing that he was "still talking in stovepipes—'wirelines,' 'mass media,' 'wireless'"—Mr. Tenhula moved on to issues affecting **Wireless Migration**:

• The process had started about 10 years before with **Personal Communications Services (PCS)**; having worked on that issue, he took great satisfaction in "see[ing] everybody walking around with these little phones."

• Already being rolled out in Europe and Asia, and on its way in the United States, was **Third Generation (3G)** or **Advanced Wireless** communications, which promised to be *the* mobile broadband platform of the ensuing few years.

• **Broadband Satellite Services** had had some "fits and starts" but might begin to fulfill their promise before long, especially to rural areas.

• An auction of **Returned TV Spectrum**, a byproduct of the media migration that related to the digital TV transition, would be undertaken by the FCC. A great deal of spectrum from TV channels 52 to 69 would be on the block, something that dropped "another hot topic," public safety and homeland security applications, into the FCC's lap. "Some would call it a pothole, some would call it a detour on this road," acknowledged Mr. Tenhula, "but it's a very important element of serving specialized enterprises with wireless communications."

- **Secondary Markets** were enhancing flexibility in the use of radio spectrum by making it possible to lease spectrum, and to trade it freely, as if it were "any other raw material."
- **Cognitive Radios** would increase spectrum sharing, since they were able to find holes in the spectrum and thereby allow the use of spectrum as available.
- **Unlicensed Devices** and **"Hot Spots"** would also demand the FCC's attention.
- **Ultrawideband Devices and Applications**, among other new technologies, had recently been provided more access to the spectrum and authorized.
- **Transitioning from Command-and-Control Regulation** of the spectrum.

POWER TO THE PEOPLE

Turning to the theme of consumer-driven innovation that Mr. Powell emblematized in the slogan "Power to the People," Mr. Tenhula began with the question: "How are the FCC's policies empowering economic opportunities and entrepreneurship?" In an address at the National Press Club earlier in 2004, the FCC Chairman had talked about getting communications and computing power "to the edges." Resting in the hands of consumers, of end users—rather than in the hands of large, centralized institutions—would be the power of the silicon chips, massive storage, and speedy connections that combine to produce smaller, more powerful devices and very exciting applications. There were already "tons" of examples of this change, including:

- digital cameras and photo printers, which had allowed photography to move from the darkroom into the home;
- iPods, MP3 players, and downloadable services, which were combining to replace CDs;
- peer-to-peer communications and file sharing;
- personal video recorders, such as the TiVo, which give users control of what they want to watch and when they want to watch it;
- private movie theaters, found in family rooms and minivans;
- GPS satellite receivers, which were becoming standard on tractors, allowing farmers to know exactly where to plant their seeds; on automobiles, allowing drivers to know their location should they get into trouble; and in cell phones, fulfilling E911 requirements that the user be locatable through the phone in an emergency; and
- WiFi VoIP phones that can bypass public networks.

Such were their economics that these devices were likely to become increasingly more powerful and less expensive, and they would require networks of networks to serve them and applications to ride on them. Consumers had been embracing them, and incumbents, in addition to entrepreneurs, were providing them. A desire to regulate them, which he attributed to "speculative fears," had raised its head.

FCC Chairman Powell believed the benefits to Americans of such innovation to be enormous, as it furnished them with more choices, better value, and more control to tailor how they communicate and get information. "Credit for these successes rests primarily with entrepreneurs," Mr. Powell had said in a speech. "But government's commitment to focus on innovation in its regulatory policies, remove unnecessary regulatory chains, place faith in the free market, and promote technology solutions has paid dividends." The FCC's objective, as described by Mr. Tenhula, was "putting the 'public' back into the 'public interest' through inevitable innovation."

WIRELESS: A MODEL FOR DEREGULATION

Looking ahead to the remaining challenges of spectrum-policy reform, he suggested that "the wireless way of getting to multiple broadband platforms," whose results he judged to be "pretty promising," offered a successful deregulatory model for the rest of telecommunications based on flexible, market-oriented regulations, as opposed to "command-and-control" restrictions of the past. Although the number of national providers of wireless telephone services had shrunk to five with the Cingular-ATT merger, the country still had multiple regional and rural providers of wireless services, as well as dozens of niche players, especially in the middleware and applications layers. Such variety meant that consumers would enjoy lower prices and more choices, and also that there would be greater innovation and deployment, with both taking place at higher speed since flexibility means that the regulator need not be consulted at every turn. Since the FCC's first auction of licenses for PCS, in 1995, subscribership had risen by 407 percent, from 28 million to more than 142 million; the percentage of the U.S. population with access to three or more providers had increased to 97 percent from 2.5 percent; and, currently, 78 percent had access to five or more wireless providers. Meanwhile, the average price per minute of services had decreased from 47 cents to 11 cents. "We've got more [flexible] spectrum in the auction pipeline," he said. He cautioned that "most [spectrum] was still under a 'command-and-control' regime" in need of its own transition.

Reforming spectrum policy was, therefore, a key item on Mr. Powell's agenda. Even as "demand for spectrum [was being] driven by an explosion of wireless technology and the ever-increasing popularity of wireless services," he had observed in late 2002, it was still under "a spectrum management regime that [was] 90 years old" rather than one "rooted in modern-day technologies and markets." Going down the lengthening list of factors that were propelling demand for wireless services, Mr. Tenhula cited the rise of the service sectors, which are "communications-intensive"; the increasing mobility of the U.S. workforce; consumers' speed in embracing the convenience and increased efficiency of wireless devices and services; technological changes that had increased the diversity of service and device offerings; and the increasing prevalence in both businesses and homes of multiple computers and wireless local area networks. At the same time, he said, technological advances opening the door to changes in spectrum policy included the increased use of digital technologies with the potential for greater throughput of information; the improvement of interference-management opportunities; and the advent of spectrum-sharing technologies, an example of which was cognitive, or 'smart' radio.

FCC SPECTRUM POLICY TASK FORCE

The FCC's Spectrum Policy Task Force (SPTF), of which Mr. Tenhula was director, issued three main recommendations in November 2002:

• Migrate from the current command-and-control regulatory model to the use both of a market-oriented, exclusive-rights model—which some would call a "property-rights" model—and of an unlicensed devices, or "commons," model.
• Implement a new paradigm for interference protection.
• Implement ways to increase access to the spectrum in all dimensions for users of both unlicensed devices and licensed spectrum.

When it came to the regulatory models, Mr. Tenhula said, the Task Force had stressed that one size does not fit all. It had, in fact, recommended striking a balance among the three general approaches in assigning spectrum-usage rights. Under the exclusive-rights model, licensees would hold exclusive yet transferable and very flexible usage rights to specified spectrum bands within defined geographic areas. These rights would be governed primarily by rules protecting users against harmful interference. Under the commons model, unlimited numbers of unlicensed users would share frequencies. Usage rights would be governed by technical standards or etiquettes for devices, and there would be no right to protection from interference. Only very limited use of the command-and-control model, the traditional regime under which the government picks the use and users of spectrum, was recommended by the Task Force, which would confine its application to the areas of public-safety, international-satellite, and broadcasting services.

MOVING TO MARKET-BASED MODELS

The consequence of using the command-and-control model, Mr. Tenhula explained, was the building up of silos. "We've got labels like 'broadcast spectrum,' 'cellular/PCS spectrum,' 'public safety spectrum,'" he said. "Folks come to the FCC begging on hand and knee: 'Can we increase power?' 'Can we have some more spectrum?' 'Can we change the service?' And the FCC takes a

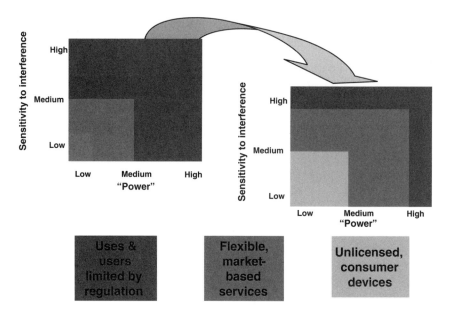

FIGURE 2 Spectrum usage models: Transition to market-based models.

very long time." The goal of the Commission under Mr. Powell was to transition to market-based models by limiting use of the command-and-control model to "high-power, high-sensitivity applications" and by moving more toward flexible, market-based services and unlicensed consumer devices (see Figure 2). The Task Force under Mr. Tenhula (and former Task Force Director Dr. Paul Kolodzy) had also concluded that increased access to spectrum could mitigate the scarcity of spectrum resources. With most "prime" spectrum already assigned, it had become increasingly difficult to find spectrum that could be made available either for new services or for the expansion of existing ones. The Task Force's recommendation was to improve access to spectrum by permitting licensees greater flexibility and to promote access in all dimensions through smarter technologies.

In the wake of the Task Force's report, the FCC had been eliminating barriers to secondary markets; designating additional spectrum for unlicensed devices; improving access to spectrum in rural areas; studying receiver interference immunity performance issues and interference temperature concepts; facilitating smarter radio technologies; and conducting service- and band-specific proceedings through which it was implementing these principles. In implementing the first of these courses of action, increasing access to spectrum through secondary markets, the Commission had:

- authorized spectrum leasing in a broad array of wireless services;
- streamlined to a single day processing of most license transfer and assignment applications;
- improved the functioning of secondary markets to facilitate access to spectrum by new technologies that make "opportunistic" use of unused spectrum; this involved a concept that it called the "private commons" and under which licensed spectrum was used for a commons-like, infrastructureless approach;
- authorized public safety-to-public safety leasing; and
- facilitated infrastructure sharing, especially in rural areas.

The SPTF also had an initiative underway in the area of research and development, in conjunction with which it was awaiting a study from the Computer Science and Telecommunications Board of the National Research Council. This SPTF report had been "inspiring a lot of study and debate around the world," Mr. Tenhula said, remarking that articles related to it were landing on his desk weekly. A parallel effort, the President's Spectrum Management Reform Initiative, was taking place within the Executive Branch; it was related to President Bush's goal of furthering broadband as well. The FCC had been collaborating with the National Telecommunications and Information Administration, a bureau of the Department of Commerce (DOC), on making more radio spectrum available for wireless broadband technologies. In July 2004, DOC had issued a pair of reports containing recommendations focused on improving spectrum management, especially within the government. Implementation of those recommendations had begun (see Figure 3).

SPECTRUM ISSUES AND OPPORTUNITIES

Mr. Tenhula, reminded the audience of the disclaimer, "the views expressed by the presenter do not necessarily reflect the views of the FCC, its Chairman, individual Commissioners, the FCC staff, or the Administration," then went down his own list of "issues and opportunities" relating to the spectrum policy, some of which reflected ideas that the SPTF had not examined. As areas of opportunity, he named:

- trends toward "layered" approaches and other regulatory frameworks;
- the question of where spectrum fit within broader discussions of telecommunications policy;
- multi-mode inter-modal broadband mania, and the question of how to define supply-and-demand problems and the relevant markets; and
- the question of whether there was a need for a new telecommunications act in an IP-based world.

Moving to a "layered" regulatory model (see Figure 4) from the stovepiped model (see Figures 5–6), a possibility that academics and some in industry had

- June 24, 2004 — Department of Commerce submitted two reports to the President that presented recommendations for developing a U.S. spectrum policy for the 21st century:
  - Report 1: "Recommendations of the Federal Government Spectrum Task Force"
  - Report 2: "Recommendations from State and Local Governments and the Private Sector Responders"

- Recommendations focused on:
  - Modernizing and improving the current spectrum management system
  - Establishing incentives for achieving improved efficiencies in spectrum use and for providing incumbent users more certainty of protection from unacceptable interference
  - Promoting timely implementation of new technologies and services while preserving national and homeland security, enabling public safety, and encouraging scientific research
  - Developing means to address spectrum needs of critical governmental missions

FIGURE 3 President's Spectrum Management Reform Initiative.

- Layered model(s) should present an optimal regulatory framework for analyzing communications policy issues.

- Policymakers need to separate (or "de-laminate") the other layers (especially the services and content layers) from the physical layer(s) in order to make rational decisions in a converged (and continuously converging) world.

FIGURE 4 Issues and opportunities: Vertical silos to horizontal layers—I.

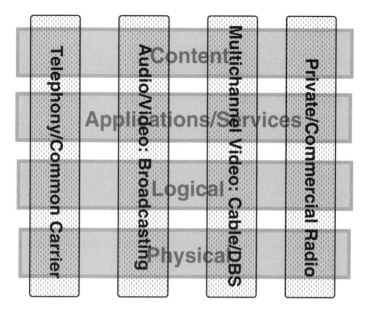

FIGURE 5 Issues and opportunities: Vertical silos to horizontal layers—II.

Where's the "Spectrum?"

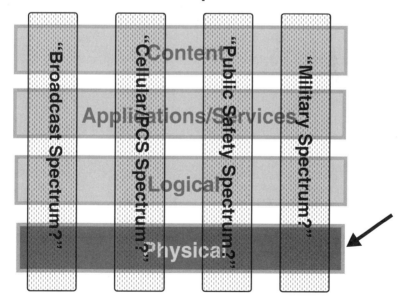

FIGURE 6 Issues and opportunities: Vertical silos to horizontal layers—III.

been exploring and that had entered into several FCC proceedings, would involve translating to a model of telecommunications regulation the Open System Interconnect [OSI] stack or the IP layers approach in the Internet. Mr. Tenhula, acknowledging a personal preference for the layered model, asserted that it could present the optimal regulatory framework for analyzing communications policy issues because it would allow policymakers to separate the layers and thereby focus decision making at the level of the problem. The current focus, he noted, was on problems at the physical layer, especially in the last mile.

Rather than preserve the artificial vertical integration that had existed for decades and had created the silos that grew up over the years, Mr. Tenhula suggested, it would make "total sense" to let the natural layers fall as they might. Replacing sector-specific telecom regulation, a "specialized, *ex ante* regime that can't keep up with technology," with a regime whose layers were not tailored to communications would be in keeping with the "New Economy": The notion that "it's all networked" would parallel that of "it's just commerce." In fact, a schematic depiction of such a layered model (see Figure 7) resembled a similar depiction of basic industrial organization (see Figure 8)—in the latter, the analogue to the IP stack is the sequence leading from raw materials, through production and distribution, and on to end-product services and the consumer interface. "I don't go to 'restaurants,'" he said, displacing the metaphor for

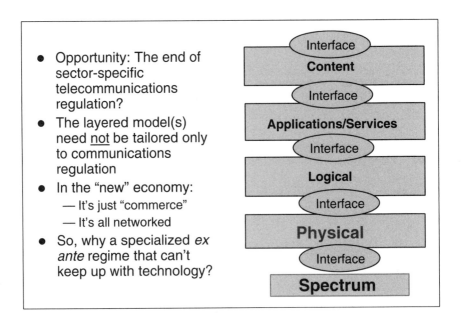

FIGURE 7 Issues and opportunities: A generic, commerce-based layered model—I.

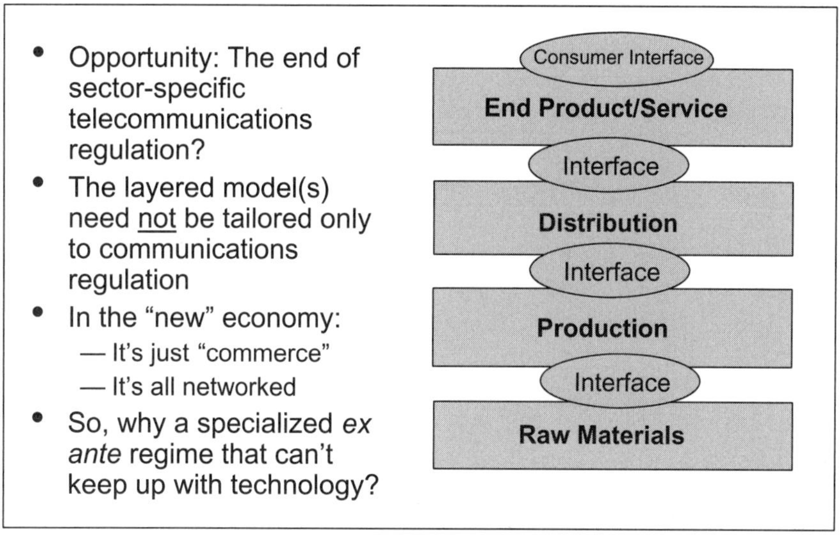

- Opportunity: The end of sector-specific telecommunications regulation?
- The layered model(s) need <u>not</u> be tailored only to communications regulation
- In the "new" economy:
  - It's just "commerce"
  - It's all networked
- So, why a specialized *ex ante* regime that can't keep up with technology?

Consumer Interface

End Product/Service

Interface

Distribution

Interface

Production

Interface

Raw Materials

FIGURE 8 Issues and opportunities: A generic, commerce-based layered model—II.

purposes of illustration, "I go to a 'food-area network'" through which the bits are 'bites' at the consumer-interface level, having gone through many commercial 'networks' of producers and distributors to get to the table.

THE UBIQUITOUS NATURE OF SPECTRUM

Advancing another argument for revising the regulatory regime, Mr. Tenhula stated that in his view "spectrum is not only in the air, but everywhere: in twisted-copper pairs, in coaxial cable, in fiber cables in the form of colors, and in powerlines." DSL is the high-frequency part of the twisted pair, coaxial cable "some video channels that are set aside from spectrum for cable-modem service." In support of this perspective, he posed the question: "Should the wireless spectrum be privatized like wireline spectrum has always been?"

Other opportunities for studying alternative regulatory frameworks included:

• monitoring European Union electronic communications directives and their implementation;

• weighing enforcement against rulemaking models of regulation—for example, the anti-trust vs. the quasi-legislative approach taken by regulatory agencies;

• reexamining the role of U.S. federalism in communications policy, something already in progress;

• considering more industry self-regulation, as well as the use of technical standards, to substitute for regulation; and

• asking whether sovereign nations and economic regions were better laboratories in a global economy, a question that recalled to him a Supreme Court dissent from the 1930s in which Justice Brandeis suggested using the states as labs.

Mr. Tenhula's suggestions for avoiding "the pitfalls of the past" in the coming broadband, IP-based world were:

• "Focus on consumer welfare, the public interest, instead of producer welfare, the special interest.

• "Think long term, and think globally.

• "Let them build it and see if they come—and so what if they don't, because others will meet that unmet demand.

• "Recognize that killer apps will come and go, and doubt all predictions.

• "Don't get bogged down in potential collateral problems like CALEA, E911, and Universal Service. Just solve them, because true innovators—like those who started MCI to compete in long-distance—won't wait, and the market won't wait, either.

• "Allow for planning. Providing time for old regulations to phase out and new regulations to phase in has worked well in certain areas where regulation is necessary.

• "Experiment on a case-by-case basis, and allow for failure."

AN UNCERTAIN LEGISLATIVE FUTURE

On the question of whether a new Telecommunications Act might be on the horizon, Mr. Tenhula deferred, noting that bringing the 1996 Act to passage took more than a decade. Comparing the approaches of 1934 and 1996 acts, he described the former as providing "very broad delegation of authority; flexible standards; little determinacy; and permissive authority for the FCC [based on] trust in the agency as an enlightened group of individuals acting in the public interest." The 1996 Act, he said, was characterized by "a comprehensive blue-print based on current technology; very prescriptive terms inviting litigation; very parochial, special interests; restraint on the regulator; and the hope that competitive market forces will someday prevail."

Was there something better than the status quo? Should the FCC become a Federal Communications *Corporation* along the lines of Fannie Mae and Freddie Mac? Should rules be abandoned in favor of an experimental expert tribunal that would resolve disputes? Mr. Tenhula predicted that a philosophical dispute, that pitting free markets against industrial policy, would be determinant.

His "personal pitch," which reflected his self-admitted obsession with spectrum policy reform, was for the passage of a new act addressing spectrum management problems. Noting that Title III of the Communications Act, which dates to the period 1912–1927, contains its oldest provisions, Mr. Tenhula set forth a potential framework for new legislation that would:

- end the government monopoly over spectrum allocation;
- end implicit subsidies and eliminate windfalls;
- allow for natural separation of the layers and sub-layers;
- provide tools for policing anti-competitive denial of access to both the wireless and wireline spectrum that, however, would not deny the benefits of efficient vertical integration; and
- maintain FCC jurisdiction over the service layers and protect consumers in cases of market failure.

Under such legislation, he asserted, the FCC might be able to act as an expert tribunal to resolve actual disputes in a timely manner.

Leaving the podium, Mr. Tenhula thanked the audience and offered the following valediction: "I'll see you on the road to the Great Digital Migration, where spectrum policy will rule."

DISCUSSION

Beginning the discussion period, Dr. Jorgenson explained that not much time remained before the scheduled break and said that he would therefore enforce the "famous White House Rule": a limit on questions of one per person. He then recognized the first questioner.

Asked by Hugh McElrath of the Office of Naval Intelligence to talk about the use of spectrum by the military, Mr. Tenhula expressed his opinion that, although the military reservation of spectrum had "been there since the beginning," it was in need of reexamination just like every other aspect of spectrum policy. And this reexamination was, in fact, taking place: A task force working under the President's Spectrum Management Reform Initiative was looking into improving efficiencies for spectrum management within the government. The White House could be expected to issue a memorandum soon, he said; it would direct agencies to do more efficient planning but would stop short of applying market principles through imposing fees or allowing the trading of spectrum. Thus, spectrum would probably continue to be, for the foreseeable future, a "unique [case] where a raw or natural resource is dedicated specifically in chunks for military or government use." Because such change did not appear likely anytime soon, other spectrum bands were under review, and it was spectrum that had been moved from the government to the commercial side that was making way for third-generation services.

John Gardinier, who identified himself as retired, observed that while Moore's Law was in force for processing, and storage capability was growing even faster, communications was really about input/output (I/O). "I can't find any way to speed up the video streams that are sometimes just not giving me the information as fast as I want to process it," he remarked, asking whether I/O rates could be improved so as to take better advantage of available technologies.

I/O Restraints Financial, Not Technical

Dr. Raduchel responded that telecommunications is in the end all silicon, meaning all computers, and that there was no technological constraint on moving information more rapidly, as "what can be put over the spectrum is driven by Moore's Law." The issues that Mr. Gardinier had raised were instead related to provisioning decisions made by network operators and to cost decisions made by users regarding the speed of the microprocessors they wanted put into their devices. "The real limitation and constraint on the Information Age" was, he predicted, going to be battery life: "It's clear that a cell phone that needs to be recharged during the day is of no use to consumers and they won't buy it." It was because batteries sufficiently powerful to last through a day did not yet exist that so much research had been going into fuel cells. He believed, however, that a way would be found to get the information to Mr. Gardinier at the speed he demanded.

Richard Lempert of the National Science Foundation said he had viewed the thinking behind Mr. Tenhula's presentation as focusing on efficiency, technology, and fairness as to who would get access to spectrum. While acknowledging the foregoing as "terribly important and understandable values," he pointed to that moment in the 2004 presidential campaign when the owner of 61 TV stations scheduled for broadcast as a documentary "what some people thought was a commercial." Mr. Lempert also related having read that "due to threats from the FCC," *Saving Private Ryan* was not shown on many networks. "Once one has concentration, which the auction method allows," he argued, "one also opens things up to a cutting back of the diversity of views that can be presented." He asked to what extent these issues, which he called "neither economic nor technological," might be figuring in the Commission's deliberations.

Mr. Tenhula recalled his personal reaction to the two controversies that Mr. Lempert had mentioned as: "There's 10 years more of the broadcast silo." The issue tied in the content layer, the very top layer of his model, with the physical layer of spectrum at the bottom. While "broadcast spectrum" is simply a means of distributing video signals, he said, because of the traditional definition of broadcasting such reactions would continue—until, perhaps, "viewership tips over" to cell phones, to cable, or to the Internet. "What it does for me is it just perpetuates those old models in the regulation of the stovepipe," he said. "It matters who has the spectrum, and then the content riding on top of that spectrum will be regulated accordingly."

South Korea Reaches a Tipping Point

Dr. Raduchel interjected that South Korea had become the first "major society" to have reached a "very interesting" tipping point: In 2004, daily Internet usage there began to exceed daily television viewing for the society as a whole. This would happen elsewhere, he predicted.

Mark Myers of the University of Pennsylvania's Wharton School said that the move from a stovepiped to a horizontal regulatory model called to his mind how the computing model went from vertical integrated systems to horizontal. But the latter model, he said, had begun to break up due to technology. "Are such models potentially limiting how we think and talk together about convergence as we go towards the future?" he asked, speculating that communications' convergence with computing could jump to a "yet-unpictured" model.

Dr. Raduchel said he thought the layered model was becoming dominant. "I see open source as the economy's response to how you build integrated products and sustain a layered model of technology," he stated. While observing that the final packaging might be changing as costs dropped and consumers wanted more complexity, he called the layered model "alive and well," adding: "If anything, it's being more sustained by all the technology developments."

Market Model vs. Legacy Claims

Jim Snider of the New America Foundation commented that while the FCC had talked "about things like the importance of broadband Internet in rural areas and flexibility," its TV translator decision of the previous September 30 would tie up hundreds of additional megahertz of prime spectrum for what he called an obsolete application. Asserting that "the translators got what the high-power broadcasters got in 1996 and 1997, which was digital flexibility and the option of a second channel," he asked how the FCC squared this result with its stated goals.

Mr. Tenhula, at pains to point out that as a staffer he did not have a vote, stated that digging into specific legacy uses and applications or into other specific problems would lead to the realization that they were secondary. With that realization would come the conclusion that they would "have to move out of the way if the new guy comes," he said. And although "they will move out of the way," the case cited by Mr. Snider had demonstrated that "they're there until somebody else wants to use that spectrum." He expected to hear more about the kind of transaction costs the decision would entail, especially in the vacated analog bands.

Dr. Jorgenson closed this initial part of the program by thanking speakers and audience alike for their presentations and questions, which he said had combined to provide "an excellent overview" of the day's subject.

New Technology Trends and Implications

INTRODUCTION

Mark B. Myers
The Wharton School
University of Pennsylvania

Dr. Myers opened the session by announcing that each presenter would have 20 minutes and that the question period would come after all had spoken. He then introduced the panel's first speaker, Mark Doms of the Federal Reserve Bank of San Francisco.

THE RECORD TO DATE:
QUALITY-ADJUSTED PRICES FOR EQUIPMENT

Mark E. Doms
Federal Reserve Bank of San Francisco

Dr. Doms, thanking Dr. Jorgenson for the invitation and Dr. Wessner for organizing the conference, began by noting that he would use the terms "technological change" and "technological advances" interchangeably. At the outset, he would tackle the question of why the link between technological change and prices is significant, so that he would then be able to explain how the advent of "some new gee-whiz technology" translates into prices and hence into consumer welfare. Thereafter he would talk about what the current official numbers said, about what other estimates were, and about the challenges of coming up with good prices for communications equipment and services—again, for the purpose of clarifying how a society benefits from innovations of the kind under consideration at the symposium. He would close by discussing the evolution of communications services and equipment.

Why do we care about prices? And why do we care about investment in communications? First, investment in communications has been substantial, but also very volatile, in the United States. As Dr. Jorgenson had said, one of the first-step products in the New Economy had been computers, but it was also true that the country's investment in communications equipment had been of about the same dollar magnitude as its investment in computers. From a GDP or national-accounts perspective, therefore, the two were pretty similar over the course of the 1990s and had continued to be similar in the current decade. Around $100 billion per year, representing a little over 10 percent of total equipment investment in the U.S. economy, was being spent on communications; Dr. Doms termed that "a fairly sizable chunk."

At the same time, there had been huge swings in the U.S. investment in communications, making it one of the most volatile of all the components of GDP. During the past recession, investment in communications gear fell 35 percent from peak to trough, "just a very, very large number." As more years of data came in, making possible a backward glance, the recession of the early 2000s might be remembered as a "high-tech recession," Dr. Doms speculated, adding that "certainly what happened to communications played a major role in what happened in the high-tech sector."

Communications Investment and National Economic Performance

The other reason for interest in communications investment springs from the way in which it contributes to the performance of the U.S. economy. Those economists who monitor the national economy, whether they work for a statistical agency or for the Federal Reserve, look at measures of how many dollars are spent on communications in the United States every year. What makes their job hard is that a dollar spent today on communications is not the same as a dollar spent yesterday; in fact, there is a great deal of change. Dr. Doms observed that a computer costing $1,000 currently was a lot more powerful and a lot more useful than a computer that that had cost $1,000 five or ten years before. The same was thought to hold true of communications gear; there had been enormous technological change, especially going back 25 years. At that time, most communications was done by landline telephone, a stark contrast to the diversity of means of communication currently available. So the problem that must be surmounted in order to understand how communications affects GDP and productivity growth is translating a given amount that was spent in a past year into today's dollars. "We basically try to ask this question: If we have $100 billion today, what did that translate to in spending, say, four years ago?"

Economists are able to look at such trends over time—to make "intertemporal comparisons"—by using price indexes. To illustrate, Dr. Doms turned to the technological change that occurred in fiber optics between 1996 and 2001. During that period, there were tremendous advances in the amount of information that could travel down a strand of glass fiber, owing to increases both in the number of channels—that is, in the number of wavelengths that could be transmitted along a single fiber—and in the capacity of each channel. Depending on how this change is measured, on the point at which the measurement begins, and so on, "you basically get a doubling every year in the potential capacity of a single strand of glass fiber," he said. During this five-year period, the price of the gear used to transmit information over fiber fell, on average, 14.9 percent per year.

Pointing out that the latter rate was clearly below the rate of increase in capacity, Dr. Doms underlined the importance of the lack of a one-to-one relationship between the change in technological capability and the price. "The intuition is that if you have the option to buy a car that's twice as fast as your current car, you will not value that new, fast car twice as much as your old car," he said, because a form of diminishing returns sets in. In a similar way, the price acts as an indicator of the value that society places on a technological change. It was unfortunately probable, therefore, that the price indexes currently in use for looking at productivity and at GDP understated the true price declines that had occurred for communications equipment.

Prices for Computers vs. Communications Gear

According to the U.S. Bureau of Economic Analysis (BEA), whose information on prices comes mainly from the U.S. Bureau of Labor Statistics, prices for communications gear fell an average of 3.2 percent per year between 1994 and 2000. That stood in sharp contrast to what had happened to computer prices, which fell an average of 19.3 percent during the same years. "With all the innovations that happened in communications equipment, do we really think that the official number of 3.2 percent per year is accurate during this time period?" Dr. Doms asked, answering: "Probably not." Work done by him and others indicated, rather, that communications equipment prices fell on the order of 8 to 10 percent, about half as fast as prices for computers. This movement stood in contrast to that of most other prices in an economy where prices tend to go up, and which was then showing an inflation rate of between 1 and 2.5 percent.

Dr. Doms then turned his attention to the challenges of measuring prices. Although made using "traditional, standard methods and crude data," the estimates of 8 to 10 percent per year through 2000 for the drop in prices of communications gear represented a step in the right direction. Still, he acknowledged, more refinement was in order on that front. Additionally, it appeared that no one had a very clear idea of what had happened with regard to technological change in and prices of communications equipment from 2001 on. Computing such prices in order to see how much better off society was as a result of the technological changes was a very hard job demanding a large number of person-hours. "We had to purchase a lot of private-sector data," he recalled, describing the undertaking as "very expensive" and pointing out that statistical agencies such as BEA and the National Bureau of Economic Research were "very budget-constrained."

An important challenge, looking into the future, was in the very speed at which technology was changing. "We don't know what technology is going to emerge three months from now, a year from now, two years from now," Dr. Doms remarked, "and it's very hard for the statistical agencies to figure out what they should be following." Was WiFi going to take off any more than it had to date—was that going to be "the next great thing"? Just how quickly would fiber to the home take off—and what would be the effect on the equipment involved in that? Those studying the economy's welfare would like to know what is going to happen in the future so that they can start gathering the appropriate data, do the appropriate analysis, and construct the price indexes.

Communications Prices Past, Present, and Future

To illustrate the increasing difficulty of tracking both prices and technological change, Dr. Doms displayed a table comparing the landscape for communications and communications equipment 25 years earlier, currently, and in the future (see Figure 9). A quarter-century back, when most of the money spent on telecommunications equipment went to switches for telephone centers, the industry was "a lot easier" to track: "We could see what happened when we went to digital switches." In the 1990s and into 2000, there was a movement away from spending on telephone switches and toward spending on a wide array of telecommunications technologies, in particular those connected to data, computer networking, and fiber optics. Following these developments was harder for the statistical agencies, especially in light of their budgetary problems. "Unless the statistical agencies get increased funding," he added, "in the future they are not going to be able to follow new, evolving trends very well."

Summarizing, Dr. Doms said that his efforts were aimed at improving understanding of how technology increases the economic performance of the country. The real terms in which GDP and productivity growth were discussed, he noted, were a tool used to control for what was happening to prices in the economy. The area of communications equipment and services was one in which, he believed, prices were "very much mis-measured," and hence this area's contribution to national economic performance was probably greatly understated.

	25 years ago	Today	Future
Primary form of communication	Land line voice, some data	Data, cellular voice, landline voice, cable	Data, video, ???
Major expenditure categories for communications gear	Telephone switching equipment	Computer network, fiber optic, telephone switching, cellular, cable TV	Computer network, last-mile solutions, wireless, ???
Primary makers of the gear	Ma Bell, Nortel	Lucent, Motorola, Cisco, Nortel, Broadcomm, Juniper	???

FIGURE 9 Evolution of communications and communications equipment.

TECHNOLOGY TRENDS, EMERGING STANDARDS, AND THEIR IMPACT

Jeffrey M. Jaffe
Lucent Technologies

Saying he planned to talk about networking technologies, Dr. Jaffe indicated that he would focus on areas where issues concerning standards and lack of clarity in regulatory policy were retarding progress. In some of these areas, the whole world was being held back, while in others the United States was being held back relative to the rest of the world. He hoped that exploring some of them would stimulate discussion, which might in turn lead to forward movement. He complimented Dr. Raduchel on his presentation, which he said made clear that the dramatic changes that had taken place in communications over the previous couple of decades would continue for the next decade or two. He reiterated that he would focus on the issue of networking, which he judged "probably more complex than some of the end-user individual aspects." Policy initiatives, he added, needed to keep pace with the rapidly evolving technological realities.

Explaining why he had labeled the changes "dramatic," Dr. Jaffe pointed out that the voice network of the future would run over the Internet Protocol (IP). Since this technology has a different capability when it comes to voice quality, cost, and reliability, this would be a major change, he predicted. More remarkable changes were on the way. For example, developments in sensor networks and personal networks could mean that cell phones would soon to be sold with a built-in camcorder. "Think of having hundreds of millions of video broadcasters—basically everybody—broadcasting to the grandparents whatever is going on in their grandchildren's life, or personal video conferencing: Think of the demand of all that on the network." In addition to commercial impacts, there would be national-security impacts, he added, because sensor networks were also critical for homeland security. All these video cameras and sensor networks, he asserted, raised important issues concerning privacy and personal liberty. This means that the right standards and policy initiatives need to deal with not only the positive potentials of the new technologies, but also some of the potential downsides.

While wireless itself was "an absolutely wonderful technology," Dr. Jaffe said, its potential is obscured because there are numerous standards for it. Among these available standards are 3G1X, EVDO, EVDV, UMTS, HSDPA, 802.11, 802.15, 802.16, 802.20, 802.21, and OFDM, in addition to public-safety standards. The resulting confusion creates challenges both from a policy perspective and from the point of view of interoperability, he noted.

Contrasting Paces of Technology Development, Regulation

Meanwhile, telecommunications services were increasingly becoming blended, with voice, data, video—all media—becoming the same from the point of view of the technology. In response, the Third-Generation Partnership Program, one of the standards organizations, developed. a standard—IP Multimedia Subsystem (IMS)—for dealing with these blended services and converged access. All these advances in networking and services were taking place in a regulatory environment that was increasingly concerned about infrastructure protection, disaster recovery, and emergency services. As a vendor, one of the things that Lucent worried about, he said, was its need to develop new products and to recognize new regulatory imperatives at a time when the regulatory imperatives were very slow to come out. "The innovators are getting out there with the innovations," Dr. Jaffe noted, expressing concern about the cost of retrofitting regulatory disciplines that are later applied on the system.

Against this background, Dr. Jaffe proposed to talk about six areas where he believed regulatory and standards issues appeared to be standing in our way:

1. **Voice over IP (VoIP).**
2. **The new IMS services.**
3. **National Emergency Planning/First-Responder Networks.** Major communications deficiencies in U.S. first-responder networks were discovered on 9/11.
4. **FTTx.** Fiber to the home or premises.
5. **Government research funding for telecom.** The industry's movement to a horizontal model and elimination of stovepipes (as it implemented the innovations of the past 30 or 40 years) had been extremely efficient for the consumer. But, from a research perspective, a major issue had arisen: Who was planting the

new seed corn for tomorrow? "That is something which I think that we as a country need to be concerned about," he stated.

6. **Spectrum policy.**

With the Third-Generation Partnership Program standardizing IMS, the voice-over-IP system was being built out. Inside the network was a sophisticated set of systems that handled the media control: Session Initiation Protocol (SIP), which was the signaling protocol for voice over IP, plus a variety of controllers, media servers, and application gateways (see Figure 10). Emergency planning, CALEA, disaster recovery, and E911 need to work seamlessly in this new environment.

Turning to security issues, he raised the problem of "Spam over Internet Telephony (SPIT)." He warned while many in the audience may not have heard this term yet, they would likely make its acquaintance soon. This was because opening up a network to the Internet Protocol meant opening it up to misuse, "something that," he said, "we need to be concerned about from a regulatory point of view." He also called protocol diversity an issue, pointing out that those most expert in the signaling protocol for the next-generation network, IP, were

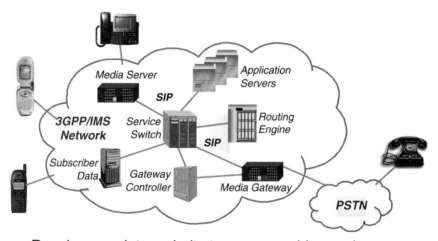

- Requires regulatory clarity to ensure no hiccups in deployment
  - Issues include Emergency Planning, CALEA, Disaster recovery
  - Security issues: SPIT, SPAM, Authentication, Denial of service
  - Protocol diversity

FIGURE 10 3GPP/IMS provides next generation of blended services and VoIP.

hackers. This was markedly at odds with the case of traditional voice networks, whose greatest experts comprised a closed society. While it is a good thing that we've opened it up, he said, we also need to have a thoughtful approach to dealing with the security issues.

Authentication: Left out of the Internet's Design

As an example of an approach that qualified as thoughtful, Dr. Jaffe cited the deliberation that led the European Union to decide that authentication was among its biggest issues. As authentication was not part of the Internet's design, with the Internet protocols it was easy for people to hide their identity. While the U.S. "Do Not Call" list had curbed telemarketing over the traditional voice network, 81 percent of email today is spam, he said. And with the simultaneous occurrence of three things—the addition of voice over IP, mobility, and SPIT—it would "be very easy for a user to spam every single cell phone in America with an SMS [Short Message Service] message." This, he noted, is "not a great thing to have in your network."

No one yet knew how to prevent this, and it would be particularly burdensome to those who had to pay when they received messages, as they would have to start paying for spam. Rather than decrying the technology, he said, he was pointing out the need to address the relevant policy issues and thereby to prevent such things from occurring. A variety of solutions to authentication—single sign-on, caller ID, public key infrastructure—had been available for some time, but their implementation had been very slow in terms of the reliability and the security of the backbone.

Similarly, he said, it was important to start thinking about how to do the signaling. For example, should there be some out-of-band signaling even within the Internet Protocol? Should giving signaling packets priority over media packets be introduced? Were there ways of introducing diversity?

Network Capability and Privacy: A Trade-off?

Dr. Jaffe then enlarged his discussion of the basic voice network to consider the services that allow the viewing of TV programs on cell phones, as mentioned by Dr. Raduchel. To be able to deliver these "lifestyle" services well, he said, IMS was designing an approach that would feature:

- seamless control;
- data transparency, meaning that everything works even with different protocols and devices;
- immediacy, in that the user is always on the network; and
- nimbleness.

By its nature, however, this network would "understand" everything about the user, which in turn raises an important social issue: If the network knows so much about end users, what does that mean for privacy? Although this is a difficult problem, he said, "I think it is absolutely vital that we address it."

Dr. Jaffe discussed U.S. and European privacy models. The U.S. model allows customers to trade off their privacy for enhanced service. The European model featured very strict laws governing the gathering and sharing of personal data. Most notable in the U.S.'s market model was the sharing of responsibilities: The government provided an overall architecture defining roles and responsibilities for network operators, network vendors, users, and so on; operators needed to obey those policies, and network vendors needed to provide technology to make it easy for users to specify their choices. He cited IBM's Hippocratic Database and Bell Labs' Privacy Conscious Framework as examples of vendors' efforts to fill the vacuum by defining approaches that allowed users to customize their privacy.

Improving Readiness for Physical, Cyber Attacks

Moving to the topic of emergency planning and first responders, Dr. Jaffe pointed to the recommendation of the 9/11 Commission that the nation be prepared to deal with simultaneous physical and cyber attacks. Also needed, he said, were trusted networks and trusted devices that could keep the government functioning in the event of emergency, and which could help first responders, whose numbers, according to some scenarios, might reach 5 million. "What we saw on 9/11 was that first responders couldn't communicate with each other if they were from different services, and that was within a single city," he recalled. "Contrast that with the broadband mobile communications capability which is available commercially." What is currently available to protect the population is a system that is not interoperable and which has very low bandwidth. By contrast, current commercial systems provide total interoperability and very high bandwidth, that offers real-time voice, video, and location services.

This led him to put forward a "modest proposal": Make available to first responders the very low cost, very efficient system of wireless communications that had been developed for commercial needs and was principally provided by cellular vendors. In our federal system, such decisions are delegated to first responders in each locality, limiting the potential for a national interoperable system. That was a problem, he suggested, that the FCC might want to take on.

Emergency planning networks were going to have to evolve so that first responders could handle not only voice networks, currently their main function, but also sensor networks. The latter, networks of highly integrated micro-sensors, would be able to provide much useful information about potential physical attacks on infrastructure. While there is a great deal of technology going into developing the sensors, arriving at the right standards for taking all the sensor information and feeding it into a national first-responder network is hard. "We desperately need standards," he stated, "and the country is not moving quickly enough."

On the subject of fiber to the home, Dr. Jaffe remarked that residential use of bandwidth had been on the same exponential curve for a long time (see Figure 11). The presence of an "insatiable appetite for bandwidth" had become apparent over the previous 20 years: "No matter how much broadband we give to people, they're willing to use it up." Despite steady improvement in modems, narrowband had reached their limits. Various forms of broadband had replaced it, but there seemed little reason to believe that demand for bandwidth would not continue to increase. This provided an inducement to achieve the highest bandwidth over the longest distances, and—despite the virtues of DSL, cable-modem, and copper—there was no question that the best bit rate for distance was provided by fiber (see Figure 12).

Improving Broadband Capacity

But leaving aside fiber for a moment, Dr. Jaffe pointed to the major challenge the United States faces in broadband: The country had fallen to eleventh in the world—or perhaps, as his fellow panelist David Isenberg interjected, to fifteenth—in broadband capacity per capita. Building costs were a factor in this, he said, but there was another reason: While fiber tended to go into new construction, regulatory uncertainty appeared to be holding down the installation of fiber all the way

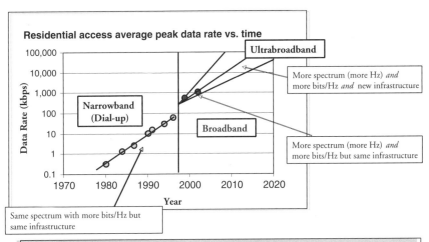

FIGURE 11 Residential access—Perspective.

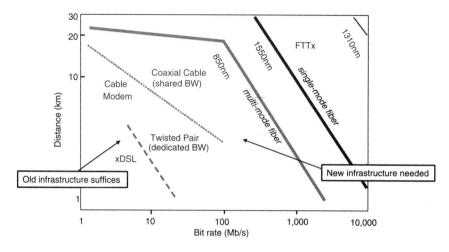

FIGURE 12 Fiber infrastructure is the enabler.
NOTE: Plot assumes modulation at 1 bit/Hz.

to the curb in conjunction with rehabilitation of existing buildings (see Figure 13). In view of this, Dr. Jaffe made a second proposal: that there be clear, consistent regulation encouraging near-term investment in a fiber infrastructure.

The lack of regulatory clarity, and by the accompanying uncertainty, erodes some of our technology leadership. For purposes of illustration, Dr. Jaffe provided the audience with an explanation of a hybrid integration technology created at Bell Laboratories. Fiber optics, he began, was very expensive technology because all the components are discrete. The cost of that expensive technology "doesn't hurt you that much" when used in a metropolitan network or in a long-haul network, because it is being amortized against billions of usages. But using that same technology at the home is very expensive because the amortization is lost. Bell Labs had developed technology in which optical components were put onto silicon wafers, thereby achieving the efficiency of Moore's Law. According to a graph on a slide he displayed, with the new technology you could substantially reduce the cost if produced in large volume. However, slow regulatory change has meant that sufficient investment in the technology has not been forthcoming.

Is Fundamental Research a Casualty of Deregulation?

Before concluding, Dr. Jaffe addressed the topic of funding for research and development, one that he said was not only very important but also close to his heart. Over the previous 25 years, the United States had substantially changed the structure of its telecommunications industry from a single, vertically-integrated company to numerous, horizontally-arranged companies. This was done for very

---

- **FTTP is the most flexible solution for wired broadband services**
  - Enormous, scalable bandwidth
  - Single-pipe integrated services (voice, data, video, ...)
  - The U.S. is generally trailing the rest of the world (11th in total broadband deployment)
- **2 scenarios:**
  - "Greenfield" -- new construction
    - Fiber all the way to the building
    - Same cost to install FTTP or copper access -- or both!
    - All new construction should have fiber installed, even if not yet used
  - "Rehab" -- rebuilding existing copper access plant
    - Small part of access network (typically < 10%) is updated each year
    - May use "deep fiber" to the curb with copper for the last 100 to 1,000 feet to building

---

FIGURE 13 FTTP landscape.

good reasons and with very good results, as it substantially introduced innovation and reduced its cost. Unfortunately, however, fundamental research had been forgotten in the process.

For many years, it was a tax on the telephone bill that had funded basic research in the telecommunications industry. After that, the venture model of the previous decade had provided a very effective substitute for underwriting research in communications: It had plowed billions of dollars, much of it focused on telecommunications, into numerous deals in the 1990s and early in 2000. Since then, however, this model had collapsed. With both models no longer relevant, Dr. Jaffe said, the United States needed a "means of cooperation across the partners in the industry to improve on the research situation." Europe, meanwhile, had adopted the explicit strategy of becoming more competitive in telecommunications and was implementing it, in part, through the European Framework Programs.

Summarizing, Dr. Jaffe called VoIP the "voice technology of the future" but reiterated numerous policy issues: security, reliability, CALEA, E911, disaster recovery, diversity, and authentication. He stressed that while new services would be enabled through the network's "knowing" a lot about the user, it was necessary to ensure that this was handled appropriately. He termed emergency planning inadequate and called for a nationwide solution based on interoperable high bandwidth. He indicated that the United States could no longer afford "to keep losing ground to the other countries of the world" in fiber to the home. And, finally, he pointed out that much of the innovation the country had seen over the previous 20 years and would see over the ensuing 10 had resulted from its past leadership in basic research in telecommunications. With this in mind, he advocated reexamining the current U.S. research model.

FOUR FUTURES FOR THE NETWORK

David S. Isenberg
Isen.com

Dr. Isenberg began by apologizing for having interrupted Dr. Jaffe, although he explained that the occasion of the interruption, an allusion to the U.S. ranking in broadband per capita, was one of his "hot-button issues." The data showing the United States to be eleventh in the world were some three years old, and in the interim this country had been growing at 42 percent per year, while a number of the countries that had placed below it in those rankings had posted annual growth rates approaching 300 percent.

The International Telecommunication Union, in a study issued early in 2004 and reflecting 2003 data, had placed the United States thirteenth. But in making his own quick analysis of the three-year-old data, Dr. Isenberg had projected that the United States would fall within a year to last among the 15 nations considered. While he had not seen a listing of countries 16 through n, he said that he would "guarantee" that some on it were growing at triple-digit rates. He would, in fact, "not be a bit surprised" to find that the United States, number three in broadband per capita as recently as 2000, had been knocked out of the top 15. "So, for all the wonderfulness of the Communications Revolution and all the improvements we're seeing in this country," he declared, "we're in a disaster: We're losing our national leadership."

Taking Intelligence out of the Network

Originally, the title of Dr. Isenberg's talk was to be "The Rise of the Stupid Network." The "stupid network" was more or less the result of applying the "end-to-end principle," which states that "if you can do something in the middle of the network or at the edge of the network, do it at the edge." Borrowing a formulation from Tim Bray, he said that the way to explain the principle to a telephone company was to say that "you want a fat pipe, you want it to be always on, and then 'get out of the way.'"

This principle—take the intelligence out of the network and put it at the edge—had guided the Internet's success. Jerome Saltzer, David Reed, and David Clark had articulated the principle, which Dr. Isenberg said was currently "the key factor," in the late 1970s. While Dr. Raduchel and others might talk about digitization and packetization as important, and while these were indeed necessary, a packetized, all-digitized network could still be a vertically integrated, stovepiped, closed network. To open up the network the end-to-end principle is needed. Indeed, he asserted, the end-to-end principle had been directly responsible for all killer applications of the previous decade, some of which he listed: email, e-commerce, Web browsing, instant messaging, blogging, audio over IP, and Internet telephony. Not one among them had been invented by a stovepiped, vertically-integrated network provider like a telephone company or a cable company. Rather, each had been brought to market as an application on top of a stupid network. The N10 network allowed future applications to be discovered.

'Any Application over Any Network'

Considering the formulation "Any Application over Any Network," Dr. Isenberg shifted focus from the first part of the phrase to the second, the "flipside," as he termed it: "Over Any Network." He presented a second list comprising twisted pairs, CoAx, Cat 5/6, fiber, hybrid fiber wireless (HFW), licensed wireless, unlicensed wireless, new wireless modulation techniques, and new wireless architectures. And, he said, more physical layers and new architectures alike remained to be discovered. Evoking the image of an hourglass, he placed the "cornucopia of applications" in the top, the Internet Protocol at the middle, and any network in the bottom.

The result, said Dr. Isenberg, is physical diversity, which avoids dependence on one set of infrastructures such as SS7. The disadvantages of such dependence were illustrated by the notorious incident in 1990 when a switch generic that was missing a semicolon caused a lengthy interruption of telephone service during which tens of millions of calls were blocked. "Physical diversity is the only route to absolute network reliability," he stated, "and you only get physical diversity with the end-to-end network." Again, it had not been telephone or cable companies that had developed the most effective of these networks: Internet, Ethernet, and unlicensed wireless. Moreover, future network technologies remained to be discovered.

Disrupting the Telco Business Model

Dr. Isenberg then offered a brief and, he said, somewhat oversimplified overview of how the N10 network disrupts the telephone company's business model (see Figure 14). Alerting the audience to what he termed the crux of his presentation, he explained: "The 'stupid' network, the end-to-end network, makes it impossible for the telephone company to sell anything—it is left with nothing to sell other than commodity connectivity." Describing the "old" model, he said that when a telco set up a call, it touched every element in every network. "This allowed the owner of Network C, for example, to introduce cool features so people would prefer it to bad old Network B, which didn't have the features." In the new, inter-networked model, it was the Internet Protocol's job to make all that was

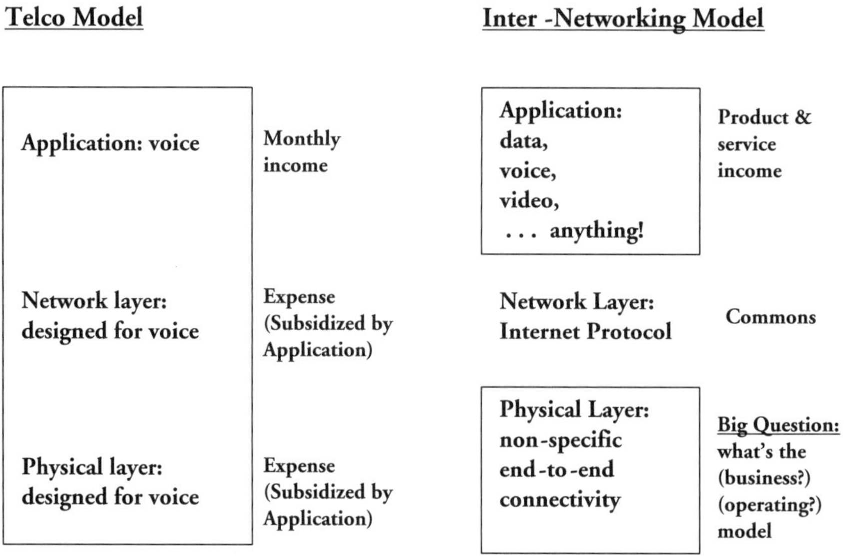

Telco Model   **Inter -Networking Model**

Telco Model		Inter-Networking Model	
Application: voice	Monthly income	Application: data, voice, video, . . . anything!	Product & service income
Network layer: designed for voice	Expense (Subsidized by Application)	Network Layer: Internet Protocol	Commons
Physical layer: designed for voice	Expense (Subsidized by Application)	Physical Layer: non-specific end-to-end connectivity	Big Question: what's the (business?) (operating?) model

FIGURE 14 How end-to-end disrupts.

specific to a single network disappear and to permit only those things common to all networks to come to the surface. Since the Internet ignores whatever is specific about a single network, including the features that had formed the basis of competition, features lose relevance in an inter-networked world.

Coming at the same point from another angle, Dr. Isenberg said that in the stovepiped, vertically-integrated model a telephone or cable company sold the application and then subsidized the underlying layers with the application revenue. In an inter-networked model users still buy the application, which still produces income; this is something the industry knows how to do. But as the applications rest upon a commons, the Internet, a "big question" remains: "What is the business or operating or functional model to get the physical connectivity?"

Future of the Network: Four Scenarios

As a prelude to sketching four scenarios for the future of the network, Dr. Isenberg characterized the status quo. The last mile was no problem anymore, as customers had 100-megabit, even gigabit LANs in their homes. Nor was the price of technology a problem, as these LANs were available for $39 at office supply stores. Rather, the current hitch was located "in the middle of the network," at the level of access (see Figure 15): "I've got a gigabit in my Macintosh sitting over there useless, because I can't connect at a gigabit."

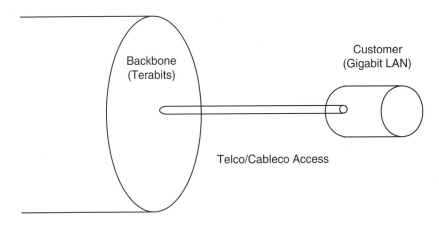

FIGURE 15 Connectivity today.

Under the first of the four scenarios, competition as envisioned by the Telecommunications Act of 1996, a variety of pipes was to go into the home (see Figure 16). The result, as described by Dr. Isenberg: "Multiple players and everybody loses." The fundamental assumption of the 1996 Act was, as he characterized it, that thanks to the market's "magic hand," competition would do what regulation could not. But he and others, among them Roxanne Googin, believed that if the middle of the network was empty and the only thing occurring there was the movement of bits back and forth, one was dealing in a pure commodity and it was very hard to have anything to sell. Once a telephone company, or a fiber company for that matter, had sold one fiber to a user, it would never sell that same user another. For, with more technology coming next year, the user would be able to light the fiber twice as fast then, and even more technology would be coming the year after that. After the initial sale, therefore, the fiber vendor would be out of business. This gave rise to a paradox: In order to survive in the competitive world of the "best network," telephone companies would have either to cripple the network or to cripple competition.

Scenario two represented the future according to the telephone companies (see Figure 17). "It is today's 'official' future," said Dr. Isenberg, "if you get around the fact that when they say 'competition,' what they really mean is 'competition where I'm the competitor.'" While there would be a modicum of improvement, far less bandwidth would be available than technology would allow or than was available in other technologically advanced countries. Besides a crippled network, this scenario featured crippled competition: Municipalities would not be allowed to compete, for example, and CLECs (competitive local exchange carriers) would have been driven out of business.

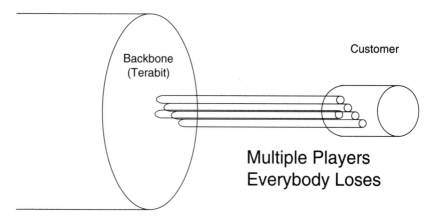

FIGURE 16 Scenario #1: Competition (as envisioned by 1996 Telecom Act).

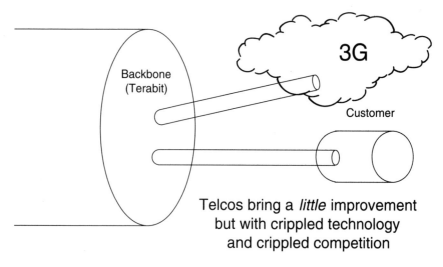

FIGURE 17 Scenario #2: Telco-topia—the "official future."

Entertaining a Forbidden Thought: Monopoly

Scenario three, which he called "rethinking 'natural monopoly,'" was "politically incorrect," Dr. Isenberg acknowledged, although he added: "what the hell." He averred that the Bell System had, for 50 years, given the United States what was arguably the world's best telephone system, albeit a vertically integrated one. He proposed, therefore, determining what the current natural monopoly was and whether something useful could be based on it (see Figure 18). Pointing out

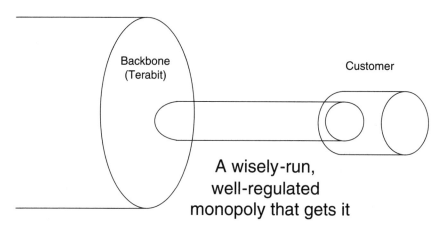

Backbone
(Terabit)

Customer

A wisely-run, well-regulated monopoly that gets it

FIGURE 18 Scenario #3: Re-regulation—rethinking "natural monopoly."

that monopolies in themselves are not illegal but only become illegal when they engage in certain behaviors, he posited that a monopoly might be crafted that was wisely regulated, well run, and public spirited.

Under the fourth scenario, technology will become so good that customers would simply build and own the network (see Figure 19). "We'll go down to the Networks 'R' Us, buy a network device, plug it in, and be on the network." There were reasons to believe, dating back to Tim Shepherd's famous MIT thesis in 1995, that this was within the scope of current technology.

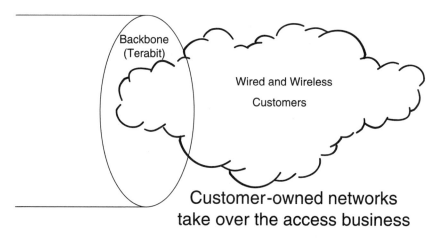

Backbone
(Terabit)

Wired and Wireless

Customers

Customer-owned networks take over the access business

FIGURE 19 Scenario #4: Customer-topia.

Speculating on what a future telecommunications act might mandate, Dr. Isenberg concluded with the following questions: "Will we be locked into an 'everybody-loses' situation, or a 'too-little-too-late' situation, where the United States loses its global leadership? Or will we manage to come up with some kind of monopoly at the very lowest layers, perhaps a monopoly that just strings fiber but doesn't light it? Or will we encourage the kind of technology whereby we don't even need a company to run our networks for us?"

DISCUSSION

Cynthia de Lorenzi introduced herself as chief executive officer of PatriotNet, an independently owned Internet service provider (ISP), and as a representative of the Washington Bureau for ISP Advocacy (WBIA), an organization made up of "the abundant small ISPs who helped grow the Internet." She asked advice on what message she might take back from the symposium to her colleagues at WBIA, to the CLECs (competitive local exchange carriers) they worked with, and to others who, she said, "help drive this industry." Voicing the claim that the small ISPs were actually the doorway to innovation, she asked what their future was and whether it was "time for all of us independents to go away."

Dr. Isenberg stated that "if nothing else gets done and the current policy directions are carried out," the industry was moving towards a reverticalization in which it would makes sense only for connectivity providers to be ISPs. Agreeing that small ISPs were "very pro-innovation," he said he shared her worry.

Dr. Jaffe, recalling technical challenges he had outlined concerning quality, security, reliability, understanding of users, providing privacy, and so on, remarked that there were many open issues in the next generation of network. Innovative service providers large and small would likely have a very important role—"introducing the necessary enhancements in an entrepreneurial way"—and he would, he said, encourage the small ISPs to look at such things as ways of providing "a better VoIP solution than the next guy." There would be a great deal of competition in that area, he predicted.

Transportation a Substitute for Bandwidth?

Jay Hellman, who introduced himself as "a real estate developer with too much of a technology education to think like one," indicated that the theme of his question would be the relationship between transportation and communication. Office buildings, which he had been in the business of constructing, and computers are not nearly as different as they appear, he postulated. Office buildings were invented as a tool for what was then a new kind of work: processing and communicating information. "The office building," he explained, "was an information processing factory, and the paradigm was paper-based manual labor." While it was a familiar fact that location is of paramount importance in real estate, he stated, "the technology that defines location is transportation." It was a point so obvious that one rarely thought about it: The easier it is to get there, the more valuable the location is.

Turning to communications, Mr. Hellman said that he had been led to start a telecommunications company by his frustration with the attempts of existing telecom companies to pass off DSL and cable modems as broadband. His conclusion, he said, had been that the last mile of the broadband network was neither copper nor CoAx; it was asphalt. "When you need bandwidth," he explained, "you get in your car and go there," stating that those in attendance had traveled to the symposium "for bandwidth." Although personal, face-to-face meetings were undeniably of value, they were overused, with the consequence that the transportation network was "in complete congestive overload." A possible factor in this was the principle enunciated by Dr. Jaffe: No matter how much bandwidth people have, they want more.

It was in light of the relationship between communication and transportation, Mr. Hellman said, that he hoped the panel might address the issue of regulation. Putting fiber into the home and making sure that it functions is a business, and it needs to be a profitable business. But, comparing it to the building of streets, he suggested that in the interest of ensuring a fair rate of return, it "ought to be a regulated business, almost like the real AT&T." Carrying the analogy further, he likened the duo of fiber and services to that of the public thoroughfare and such service companies as UPS and FedEx that use it to compete; it was desirable, he added, that the street be as accessible to as many people as possible. What did the panelists think, he asked, of the contention that the last mile of the network should be not competitive but regulated, yet that it needed to provide significant bandwidth and to be ubiquitous?

Networks in the Hands of Customers

Dr. Isenberg, acknowledging that Mr. Hellman was "onto something," named two "antidotes" to the kind of regulation he had spoken of: (1) technological decline and continuation of business as usual, and (2) the development of technologies that allow customers to own their own networks. He called Mr. Hellman a "pretty good example" of the latter, since he had started a telephone company after being unable to contract for the telecommunications services he needed. Dr. Isenberg hoped that the regulatory situation would, he told Mr. Hellman, "encourage people like you to do your own thing," adding that he was "on the right track as far as thinking about the larger, more generic solution."

But regarding another of Mr. Hellman's points, that there was no such thing as too much bandwidth, Dr. Isenberg cautioned that telephone companies would take exception. "If they served up too much bandwidth, then they wouldn't have anything to sell," he stated, arguing that "telephone companies make their profit based on scarcity."

Toward More Sophisticated Price Indexes

Dave Wasshausen of BEA's national accounts staff registered his agreement with Dr. Doms that having good price indexes for high-tech communications equipment, computers, and software, is vitally important to measuring real investment in the national accounts. While admitting that his colleagues generally used the Producer Price Index in their work, he noted that they were receptive to work on indexes being done by academics and in the private sector, and he pointed out that they had incorporated Dr. Doms's work into their price indexes for LAN equipment, switch gear, and other types of high-tech equipment. Such symposia as the present one were very important to him and his colleagues, as they wanted to learn more about how to measure such equipment. Finally, recalling Dr. Doms' allusion to the budget constraints under which statistical agencies found themselves, he agreed that BEA had to prioritize. From his own perspective, the highest priority at that moment was price indexes for software; this matter had been treated at the STEP Board's symposium of February 2004, which he planned to revisit.

Dr. Myers, returning to Dr. Jaffe's comments on the demise of basic research within corporations, observed that great basic research laboratories had been created and supported by monopolies. In addition to the AT&T and IBM monopolies, there had been a Xerox monopoly and, for a number of years, a monopoly held by DuPont. Given the contemporary consumer-oriented, market focus, he said, it was unlikely that a government monopoly would be created. "The only monopoly I envision occurring would be a 'market monopoly,'" he said, pointing to the "Wintel" monopoly reigning in personal computing as an example. Expressing his doubt that this monopoly had created the kind of basic research evoked by Dr. Jaffe, he asked for the latter's comments.

New Models for Basic Research

Dr. Jaffe praised the government for doing an "outstanding job" in funding fundamental research within the university system—a role identified over half a century before. As the nation moved toward having fewer natural monopolies, he said, it needed a way of funding basic research in the commercial sector, which "brings a different perspective than the university system." That challenge was currently being studied by a panel of the National Research Council's Computer Science and Telecommunications Board, whose report was due out early in 2005. Led by Bob Lucky, the panel was to consider the dimensions of telecommunications research and, in addition, the models that might be appropriate for it.

Dr. Myers expressed the STEP Board's appreciation to the panelists.

The Broadband Opportunity: What Needs to Be Done?

INTRODUCTION

Kenneth Flamm
University of Texas at Austin

Convening the panel, Dr. Flamm offered introductory remarks focusing on the areas it was to consider. The first of these concerned the definition of broadband, and, specifically, the speed of transmission to which the term properly applied. Discussion by the previous panel of the United States' slipping behind in penetration rates for "something called, quote, broadband," he said, had failed to acknowledge that what qualified as broadband had been changing as well. Some of the countries that were pulling ahead of the United States in penetration rates were also offering higher quality, faster broadband connections. The FCC used 200 kilobits per second (kbps) as its threshold speed for broadband. He pointed out, however, that according to this definition, the variation between fast broadband and slow broadband was greater than an order of magnitude—and thus substantially wider than the variation between slow broadband and low speed dialup, since the former was only four times faster than the latter. "Increasingly," he said, "what broadband is, and the quality of broadband service, are going to be an important issue."

Broadband Access: A Rapidly Changing Landscape

A second area, broadband access, had been an issue for many years in the United States, but data that the FCC had been collecting on broadband availability had shown that the landscape was changing very rapidly and that it might no longer be an issue. At present, high-speed service was available in around 86 percent of Zip Codes containing 99 percent of the U.S. population, and approximately 40 percent of U.S. Internet households connected with broadband (see Figure 20). "So the question is," Dr. Flamm said, "if the definition of broadband is one issue, and broadband access is no longer really an issue, what are the issues?" The panel, he anticipated, would focus on pricing and on competition. The latter, as measured by the number of providers per Zip Code, had changed rapidly over the previous four years (see Figure 21). Still, there remained many Zip Codes where the traditional duopoly—made up of the telephone company and the local cable company—were the sole providers of broadband services.

He would take advantage of the occasion, Dr. Flamm said, to bring up what he called a "very important data issue" and to make a plea for funding. In 1999 and 2000, the Bureau of Labor Statistics had sponsored a survey sampling U.S.

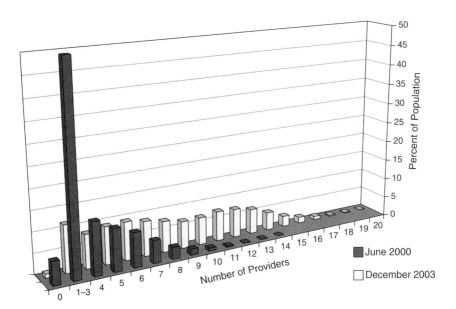

FIGURE 20 99 percent of population now has at least 1 provider in their Zip Code: Population-weighted distribution of Zip Codes by number of broadband providers.

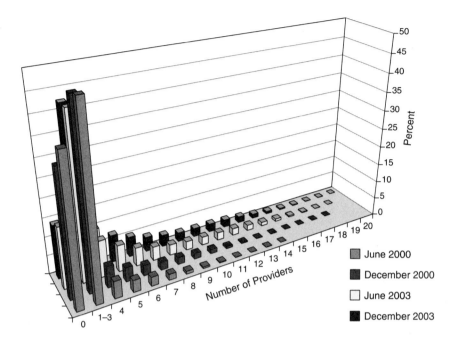

FIGURE 21 Rapid change in U.S. broadband penetration, competition over four years: Distribution of Zip Codes by number of broadband providers.
NOTE: FCC numbers have been corrected to reflect undercounting of rural Zip Codes with zero providers; corrections probably somewhat overestimate zip codes with zero providers.

households that collected data on broadband prices. The question on broadband prices had been discontinued in 2001, but why? His guess was that broadband was held to be less important then than it had been in 1999 and 2000. "There are some issues of priority in our statistical data-collection efforts," he declared, "that need to be addressed."

Dr. Flamm then introduced the panel's first speaker, Charles Ferguson, whom he billed as "interesting and often provocative." Dr. Ferguson had left MIT to become one of the founders of Vermeer Technologies, where he had participated in the creation of a software technology that had subsequently become known as FrontPage. Vermeer, like quite a few successful startups, had been acquired subsequently by Microsoft, and FrontPage was being sold as part of the Microsoft Office suite. As a result of the sale of the company, Dr. Ferguson held a fair amount of Microsoft stock; he declared, however, that he had no financial interest in any telecommunications provider.

THE BROADBAND PROBLEM:
MARKET FAILURES AND POLICY DILEMMAS

Charles H. Ferguson
The Brookings Institution

Dr. Ferguson said that his talk would address three issues:

1. the size of the problem that the United States was facing in the broadband area, which, as others had stated, was quite substantial;
2. why that problem existed; and
3, what could be done about it.

He began with two points of presumed general understanding and agreement: that a rapid de-maturing of traditional analog consumer electronics was taking place; and that, technologically at least, the potential existed for de-maturing of the analogue telecommunications industry, which had previously been dominated by telephone service. Further, he postulated that a majority in the information technology industries would substantially agree on what the "radiant future" should look like, remarking that, in a few sectors, this future had already been reached.

United States Lagging in Broadband Penetration

Dr. Ferguson pointed, however, to an "enormous gap" existing between this vision of the sector's potential and its actual state, particularly in the United States. And perhaps of foremost interest, he said, was the extent of the gap between the United States and other nations. Those who had visited China's large coastal cities in the previous two years understood the U.S. lag; the same was true for those who had been to South Korea, Japan, Taiwan, or other parts of industrialized Asia, as this gap was very obvious.

To illustrate, he posted graphics highlighting year-old data with the comment (see Figures 22 and 23): "As Mr. Isenberg mentioned a few minutes ago, things have gotten substantially worse since then." These figures showed that, as of late 2003, DSL penetration per 100 telephone lines had reached 27.7 in that category's leading nation, South Korea. The United States, which was not in the world's top ten, had only 4.8 DSL subscribers per 100 phone lines and trailed China, which was at 5.1. Moreover, when it came to the absolute number of DSL lines, both China and Japan had surpassed the United States, which, Dr. Ferguson conjectured, had likely fallen in the intervening year from third place to sixth or seventh.

Global Ranking 31 December 2003	Country	DSL Subscribers 31 December 2003	Telephone lines 2001/02 (ITU)	DSL per 100 phone lines 31 December 2003
1	South Korea	6,435,955	23,257,000	27.7
2	Taiwan	2,800,000	13,099,416	21.4
3	Hong Kong	690,000	3,842,943	18.0
4	Belgium	789,677	5,132,427	15.4
5	Japan	10,272,052	71,149,000	14.4
6	Denmark	473,481	3,739,247	12.7
7	Singapore	242,000	1,927,200	12.6
8	Israel	380,000	3,100,000	12.3
9	Finland	336,600	2,850,000	11.8
10	Canada	2,170,243	19,962,072	10.9

FIGURE 22 DSL penetration by country—I.

Global Ranking	Country	DSL Subscribers 31 December 2003	DSL per 100 phone lines 31 December 2003
1	China	10,950,000	5.1
2	Japan	10,272,052	14.4
3	USA	9,119,000	4.8
4	South Korea	6,435,955	27.7
5	Germany	4,500,000	8.4
6	France	3,262,700	9.6
7	Taiwan	2,800,000	21.4
8	Italy	2,280,000	8.3
9	Canada	2,170,243	10.9
10	UK	1,820,230	5.2

FIGURE 23 DSL penetration by country—II.

Slow Growth Holding United States Back

Accounting for this drop were growth rates. The total U.S. growth rate for DSL lines, both in absolute number and in penetration per telephone line, was something on the order of 40 percent, with the rural growth rate at about twice that. Dr. Ferguson acknowledged that the cable telephone system still provided the majority of U.S. residential broadband connections; he contended, however, that that did nothing to change the overall picture. In the first place, he explained, when business connections were included, the percentage of total U.S. broadband connections provided by the cable telephone system was relatively low. In the second place, even in the residential market the percentage of connections provided by the cable system had been holding roughly constant, as had the cable system's growth rate in respect not only to connections but also to bandwidth levels.

Price/Performance Another U.S. Weakness

In fact, although the U.S. cable television system had been improving its bandwidth levels slightly faster than had the U.S. telephone system, the price-to-performance ratio for broadband services abroad was enormously superior to that in this country. Depending on the nation and the service being compared, other nations were outpacing the United States by between 2 and 15 times. And the gap was growing rapidly, because U.S. price/performance in local telecom services—not only in digital services like DSL and local broadband, but also in voice telephony and such related services as voice mail and caller ID—was, "quite astonishingly," roughly flat. The total local telephone bill in the United States was flat or even increasing, a surprising fact in that the underlying technology for all such services is computing—which was improving, depending on the technology being measured and the measurement being used, anywhere between 20 percent and 50 percent per year. Alluding to Dr. Doms's discussion of the increase in the capacity of fiber-optic cable deriving from increases of both the number of channels and the capacity of each channel, Dr. Ferguson said the technological improvement was, in some cases, as high as 100 percent per year.

Both DSL and cable-modem service, however, had displayed very low rates of progress. On top of that, the benefit that those broadband services had provided relative to that a simple modem could provide had turned out to be, in Dr. Ferguson's words, "surprisingly modest." Modem technology had improved at a rate of around 40 percent per year until reaching its limits at 56–60 kbps; modem service, he pointed out, was "to a first approximation, symmetric." It had then taken several years for consumer DSL services to be introduced, and they had been relatively expensive upon introduction; furthermore, they are asymmetric, something he called "not accidental." The result, combining all those factors, was an annual improvement in the price/performance of bandwidth of some 10–

15 percent yielded by either DSL or cable-modem service. Not only was this rate already far below that which the technology curve should have been providing, but it seemed to be in the process of slowing even more.

Bandwidth Costs Dominant

Dr. Ferguson commented further that bandwidth rather than computer hardware frequently dominated the total cost of adoption of a new network-computing application. Personal computers were powerful and cheap, but deploying a high-performance, high-quality videoconferencing system could nonetheless prove extremely expensive. Purely for purposes of illustration, he posited the use of T-1 service, whose price/performance, he pointed out, had improved very slowly if at all over the previous several years. If two T-1 lines were required for point-to-point connections between two personal computers, over a three- or four-year period the total costs of using that service would be completely dominated by bandwidth costs.

Competition in U.S. Markets Flawed

What was the reason for this? While allowing that in various respects regulatory costs imposed on the entire system accounted for some degree of drag, this was "not the principal story" in Dr. Ferguson's opinion. Very simply, he stated, there were two monopoly industries providing broadband service in the United States, both had very severe conflicts of interest, and they avoided competing with each other except in the residential market. And even in the residential market, their competition with one another was "quite restrained, and much less substantial than you might suspect."

The conflict of interest of the telephone companies was, Dr. Ferguson asserted, "fairly obvious": They had incumbent businesses that were providing very expensive voice and traditional data services, and very rapid improvements in price/performance of bandwidth would have undercut their dominant businesses in a major way. The same was true of the cable system: It provided video services that could easily be provided over a sufficiently high-performance IP network.

Additional conflicts of interest in both industries related not only to Internet telephony but also to intellectual property rights and to proprietary intellectual property control. This was particularly so of the cable television industry, which had many proprietary-entertainment and other content assets. It was very afraid of the effects of piracy, and one consequence was that cable operators wanted to provide downstream-weighted services, because "upstream service is what determines piracy levels when you're using peer-to-peer networks," Dr. Ferguson stated. Cable-modem service, like DSL, is a very asymmetric service and heavily weighted downstream; and the reason that the telephone companies preferred downstream-weighted is that symmetric service would make it far easier to use Internet telephony.

'Local Bandwidth Bottleneck' Hurts Computing

Distortions existed in both industries, he said, and not only about service's price/performance but also about its technical characteristics, its quality levels, and the degree to which it was symmetric. He judged the economic stakes involved in this question to be "quite large" for the country. While computing them in a rigorous way would be extraordinarily difficult, it seemed increasingly clear that the "local bandwidth bottleneck" was having a substantial effect on the growth of the computer industry, of various other portions of the information technology hardware sector, and of the American economy. "You can convince yourself reasonably easily," he stated, "that this effect is something on the order of one-half of 1 percent—or even up to 1 percent—per year in lost productivity growth and GNP, which is a lot."

This obviously had some effect, although it was not clear how substantial, on American job losses, Dr. Ferguson said, alluding to the prevailing debate over outsourcing. There was no question that broadband infrastructure was having a significant effect on the way industry was growing in China and Southeast Asia. While India had traditionally been far, far behind in telecommunications infrastructure and was still far behind both the rest of Asia and the United States, even it was gaining rapidly: Although from a very, very low base, the number of broadband connections in India was going up quite rapidly, on the order of 300 percent per year.

Broadband Shortcomings Hurt National Security

In addition to the direct economic effects, there were quite significant national security effects arising from forgone opportunity and capability in the broadband system under which the United States was laboring, Dr. Ferguson said. First, any major terrorist emergency, such as an attack that used weapons of mass destruction, would undoubtedly result in major quarantines and disruption of transportation systems. It would be imperative for many people to be able to talk to each other and understand each other's concerns at high bandwidth across wide geographical distances and with impaired mobility; the utility of a nationwide broadband system in such a situation is obvious. Then, in light of recent events in the Persian Gulf, one might also ask about the impact of reduced transportation demand on oil prices, oil security, and so forth. "Once again," he said, "one can convince oneself that the issue here is really quite substantial."

To whatever extent the United States faced a problem of "digital divide"—disparity in broadband, Internet, and computing access as a function of economic ability and economic status—that problem was also coming to be increasingly dominated by the bandwidth question. The reason here, again, was that band-width dominated the total cost of adoption of new computing applications.

Can Policy Changes Help United States?

How, then, might the United States attempt to address this question, which was of such macroeconomic and military significance? The nations that were ahead of the United States, in what they had done in the broadband-policy arena, had evidenced two shared characteristics. The first was that their governments had been "much more heavily involved in providing incentives and/or money and/or direct construction of networks than is the case in the United States," Dr. Ferguson said. The second, also related to governmental policy, was that their systems were much more competitive than that of the United States. There might be many more providers, but even when there was a relatively low number, the providers were under government pressure to improve their price/performance and to compete with each other.

This was true even where there was no explicit antitrust policy. Certainly none existed in China, but the Chinese government obviously had made very clear to the country's principal telecommunications providers that broadband deployment was a major national priority and had put them under a great deal of pressure to continue accelerating it. The case was similar in Japan and Korea, and even in non-Asian countries—Canada, for instance, and the Scandinavian countries—that had surged well ahead of the United States.

Dr. Ferguson cited as a "somewhat hopeful recent development" the FCC's unanimous vote to preempt regulatory control of voice over IP, "so at least there will not be a patchwork of 50 different state regulations of Internet telephony." He charged, however, that the United States had been "notably absent from productive efforts in regard to broadband for quite some time."

RBOCs' Consolidation 'A Major Mistake'

This was not specific to the Bush administration, he said, although he gave it lower marks than its predecessor did. What had begun to undermine the potential benefits of the Telecommunications Act of 1996 in a significant way was the series of mergers among Regional Bell Operating Companies that had effectively halved their number to four.[3] "That consolidation was unopposed by the FCC and by the Justice Department," Dr. Ferguson observed, commenting: "That, I think in retrospect, was a major mistake." A great deal of litigation had followed, and there had "not been much effort by the federal government and/or the FCC—

---

[3]While consolidation among RBOCs may have eliminated a source of potential competition, the issue of whether it eliminated actual competition and whether potential competition was likely or unlikely to begin with remains to be resolved.

depending on whether you want to differentiate between the two—to ensure that there is a competitive, open-architecture system."

Nonetheless, he did not see new legislation—even if it might be helpful in providing what he called "appropriate broadband policy"—as necessary. What was required instead was "some kind of national political will." This could open the door to a range of measures, among which might be:

- subsidizing deployment of municipal networks;
- offering investment incentives to any and all providers, whether public or private;
- constructing a large federal network; and
- putting more pressure on incumbents to open up their networks so that there was an open-architecture broadband system in the United States that was more analogous to the structure of the Internet itself.

Calling this last point critical to the future of network architecture and services, he called for a "far more open-architecture, computer-industry-like structure and feel" for the U.S. telecommunications system.

Commoditization No Enemy of Investment

In closing, Dr. Ferguson questioned Mr. Isenberg's suggestion that, since the network logically anticipated for the future would function as a commodity provider of bits, industry would likely be averse to committing to sufficient levels of investment and research. The history of the computer industry did not support that proposition, he claimed, and neither did the behavior of those Asian and European nations with broadband service superior to that available in the United States. On the contrary, most sectors of the computer industry were composed of large companies that produced commodity products in brutal competition. While there were a few notable, very profitable exceptions—he named Intel and Microsoft—more typical as examples were Dell, Hewlett-Packard, most of IBM, most of the semiconductor industry, the entire disk-drive industry, the display industry, and so on.

"Disk drives are like disk drives, bits are like bits, DRAMs are like DRAMs," he stated. "And while those industries are volatile, one does not see any hesitancy for entry or investment. Indeed, competition in those sectors is quite healthy." If there were an appropriate government policy, he concluded, it would be reasonable to expect that levels of investment and technological progress in the industry would be sufficient.

Dr. Flamm, asking the audience to hold its questions, recessed the panel's proceedings until after lunch.

Resuming the session after the lunch break, Dr. Flamm introduced Mark Wegleitner, the chief technology officer of Verizon, as the only one among the panel's distinguished speakers about whom the previous presenter, Dr. Ferguson, had had many good things to say in his recently published book.

THE BROADBAND CHALLENGE: A TELECOM PERSPECTIVE

Mark A. Wegleitner
Verizon

Mr. Wegleitner expressed his appreciation at having the opportunity to speak about a subject that his company was pursuing with great vigor, spurred by a conviction apparently shared by many at the symposium: that broadband could be an engine for growth. There were two dimensions to broadband's contribution. One was in the capital investment required to build a network and the multiplier effect that could be expected to have on growth, jobs, and innovation. Obviously, however, the investment was not an end in itself; the second dimension, at least as envisioned by Verizon, was the stimulation to the economy, as well as the social good, that would come of having a broadband network available.

Although the United States had traditionally been a leader both in communications and in attracting capital, it had not been holding onto that leading position of late. Mr. Wegleitner displayed a chart based on International Telecommunication Network (ITU) data that showed the United States to be thirteenth in the world in broadband deployment. While acknowledging that there were "lots of ways to spin numbers," he stated that the conclusion to be drawn was "that we aren't leading in what we have to perceive as one of the key technologies for any national economic environment going forward."

Richer Applications in Broadband's Future

But a question underlay this last assumption: What did we think broadband was actually going to do for people? Mr. Wegleitner named applications—email, instant messaging, basic Web browsing, small file transfer—whose requirements could be met by current broadband access technologies. Such familiar applications, although important, could almost be characterized as modest. True two-way videoconferencing and gaming, as well as voice over IP, were the next step up the ladder, but they would not in themselves exercise a huge demand on broadband. Further along, however, lay multimedia Web browsing, distance learning, and telemedicine. Even beyond those would come immersive gaming and whatever means of information and entertainment delivery were to come after high-definition television: 5-megabit and 9-megabit pixel TV, 3D TV, and/or holographics.

"We know that the bandwidth demands are just going to continue to grow and grow and grow," said Mr. Wegleitner. With the richness of the future applications he had listed would come a great deal of complexity, and it was complexity that, in the world of networking, drove broadband. But while it was certain that more bandwidth would be needed over the coming two to ten years, no one could predict with great accuracy how much. Displaying a graph in which broadband requirements on the y-axis were plotted against historical time on the x-axis, he said that the middle of the scale—where cable-modem and DSL technology operated in the 3- to 5-megabit per second (Mbps) range—had been reached (see Figure 24).

Verizon Expanding Fiber to the Premises

In its fiber-to-the-premises (FTTP) deployment Verizon had set 30 Mbps as the top offering for the present but was building for 100 Mbps. "The 100-megabit bogey has been something that we've been kicking around inside Verizon for some time now," Mr. Wegleitner remarked. The company believed that fiber all the way to the home or business was the way to satisfy that, and that the home, as the "true mass-market representative," was really the key. Verizon had announced FTTP deployment in nine states, had begun it in Texas, California, and Florida, and had been building in the East Coast states. The company hoped to be in over 100 central offices by the end of 2004, and its target was to pass 1 million homes.

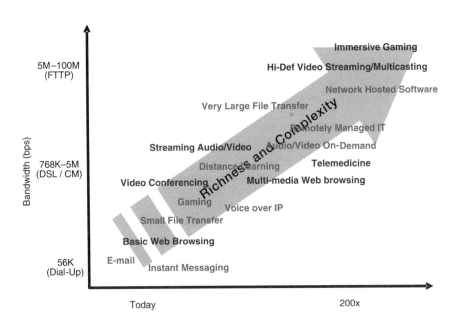

FIGURE 24 More bandwidth: Multimedia applications.

Mr. Wegleitner listed four things that can be done with data: It can be processed, stored, displayed, or moved. There had been great advances in the first three, in accordance with the well-known phenomenon of Moore's Law, as the industry had gone into more and more complex applications. But remaining to be answered was what could be provided in the form of network capability for the last, the moving of data. The objective of Verizon's business was to make sure that it was not the bandwidth bottleneck in the movement of data, which was why it had taken such steps forward as deploying fiber to the premises.

The Shape of the Internet to Come

Mr. Wegleitner next asked what the Internet would look like in 2008 and beyond. Would it remain a confederation, a descendent of the ARPANet? Or would it be a more rigorously maintained interconnection of commercial networks? His prediction was that there would always be some holdover of the Internet, but that the latter would be much more in evidence. The requirements for services offered customers would be for such quality and scope that only the interconnection of networks could provide it. Because Verizon customers would need to talk to customers of Time Warner, Comcast, or whomever else might be a provider for individual users of the network, a means of reliable interconnection would have to be available. The ability to provide service with a high level of quality would depend on an infrastructure that would go beyond what the current Internet could provide. "Maybe the Internet in itself will morph into that," he allowed, although in his own view it was more likely to be based on "interconnection agreements between commercial providers with the purpose of satisfying their own customer requirements."

Another reason this sort of commercial enterprise would underpin the country's future communications infrastructure was that a way had to be found to collect for services. The hope was that scale would drive the cost of services down dramatically and that applications, despite their increased sophistication, would be reasonably priced owing to the economics of the technology. Still, there had to be a billing system appropriate to what the providers were attempting to accomplish.

Verizon, Mr. Wegleitner explained, was building aggregation networks at the metropolitan level. It was building core network—IP backbone—nationwide, and ultimately would build or interconnect internationally as well. Motivating the company was the change taking place in the traditional telephony business; having experienced this first-hand, Verizon had seen that, to be a successful communications company, it would have to change along with the business. The $12 billion Verizon was spending annually in its capital program was, to the company's knowledge, the largest sum spent by any U.S. corporation and, perhaps, the largest in the world. "What we need to do collectively," he stated, "is ensure that Verizon puts that in places that serve not only Verizon's interests but the interests of the U.S. communications infrastructure as a whole."

Technical, Financial, Regulatory Changes

Three sorts of changes were needed to make this work. The first was technical: There were standards for particular communications protocols that needed to be taken further. One such standard, multi-protocol label switching (MPLS), had not really been developed to accomplish the commercial interconnection of networks described by Mr. Wegleitner. Equipment adhering to such standards had to become available, and it had to be deployed in the network, and agreements had to be struck among the interoperating carriers. The second was financial: A way had to be found to replace the existing "economic ecosystem," which featured access charges, a universal service fund, and other artifacts of "the way we have been operating as a communicating nation over the course of the last 50 or more years," as he put it. To move forward, a new economic system that supported the new technology deployment was needed.

The third was regulatory: While there were "any number of things" to be worked on in this arena, Verizon was concerned about "incremental rulemaking," which was the path being followed to move from the old regime to the new, with the unsuccessful result that the rules were insufficiently clear. In some cases, Mr. Wegleitner said, investments of millions or even tens of millions of dollars might hinge on the interpretation of words that, while written only a few years before, were already technically obsolete. "It's that interpretation that is going to determine the path forward of the network's evolution," he commented. Having what he called "an unnecessarily complex regulatory environment" didn't make sense in that it discouraged investment. For an example, he turned to unbundling requirements governing broadband: Although the rules had begun gaining some clarity not long before, they had earlier tamped down investment by companies and, as a consequence, scared off prospective investments. While admitting that such problems resisted simple solutions, he put forward what he called a "short answer": "Let the market rule." It was, he asserted, to a very light regulatory touch that the wireless industry owed its phenomenal growth, great innovation, and very competitive environment.

Concluding, Mr. Wegleitner said he imagined he was joining all those present in voicing the belief that broadband was the future. Telcos such as Verizon had a vested interest in broadband, as it was a key part of their future technology strategy. "We have the capability to make things happen quicker with some clarity and a cooperative effort around the technological, financial, and regulatory things that we think have to be settled going forward," he stated, adding: "The right environment here is the key."

A PARADIGM CHALLENGE:
MUNICIPALLY OWNED FIBER

H. Brian Thompson
iTown Communications

Mr. Thompson began by signaling that his views were not in accord with those of the previous speaker and by warning that some of his statements might be "less than subtle." Noting that he had joined MCI in 1981, when competition was just beginning in U.S. telecommunications, he stated that many of the issues current in 2004 had applied in the early 1980s as well. Currently, the situation in the industry was "becoming increasingly untenable," he declared.

Providing a brief overview of his talk, Mr. Thompson said that he would:

- add his own negative assessment of the U.S. competitive position to those of several previous speakers;
- discuss the essential nature of broadband;
- take up what he called "the real question," why current services were not good enough, again a topic that had received attention from other presenters;
- address the current model of the industry's fundamental structure, a matter he highlighted as "important"; and
- discuss what he characterized as the need "to make a complete change in the whole paradigm of how we're approaching the marketplace and serving the customers."

Displaying a chart illustrating broadband penetration by country that differed from that of Mr. Wegleitner only in its details, Mr. Thompson observed that the United States was far from the top of the list and sinking rapidly (see Figure 25). Concern over inequality in infrastructure and access had prompted him to participate a decade earlier in forming the Global Information Infrastructure Commission, an organization whose relevance had only increased with time, since it was no longer developing nations alone that had such problems. "We have a huge access problem in our own country," he stressed.

Broadband Central to Community Development

It was beyond question that broadband affected every aspect of a community: its economic activity, its development, its education, its delivery of health care and government services. Broadband was also the driving force behind the individual and social activity that people were willing to pay for, whether that meant information and entertainment media or "simply talking to someone somewhere else." Nations that had recognized this before it had been addressed with any effectiveness in the United States, and that had established more broadly

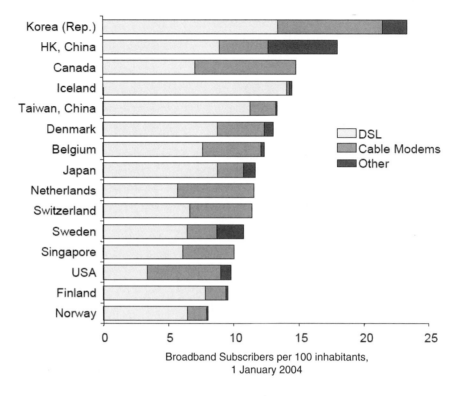

FIGURE 25 U.S. continues to lose ground on broadband penetration.
NOTE: The comparison is limited in that it does not reflect the relative prices of dial-up services in various countries or reflect what percentage of the population has access to broadband at home.
SOURCE: ITU Internet Reports, 2004.

ranging policies, had taken into account the issues of economic and community development.

Broadband Policy: An Irish Example

Mr. Thompson, an adviser to the Irish government for the previous decade, recalled putting together two study groups, the latest of which had resolved to embrace a public-private partnership at the local level in the aim of stimulating the development of home broadband access within that country. This, in turn, was expected to generate "huge" rewards in economic development, as well as to promote the entire community's technological and social development. The Irish government had chosen 19 towns of moderate size and provided funds on a revolving basis to entities that were willing to take new technologies into those towns and to establish "true" broadband—"not DSL, not simple broadband, but very high level broadband connectivity, both wireless and fiber."

The government had committed itself, he said, to moving beyond what it had done ten years before to attract the information infrastructure companies that had become a fundamental part of Ireland's economy. It had recognized that these firms were no longer making and selling shrink-wrapped software but were shipping software out over the Net. Besides addressing this change, it had established a national science foundation as a tool for staying up with the state of the art and maintaining the country's economic development. Only three or four weeks previously, Ireland's population had gone over 4 million for the first time since 1857, as people were returning to the country. Mr. Thompson likened Ireland's former condition to that of many states in the United States, which had seen residents leave for centers where they could get access to communications and infrastructure.

Why Aren't Current Services 'Good Enough'?

He then turned to the question, also raised by preceding speakers, of why current services were not "good enough." For more than five years, the national and international backbones had had a capacity of 1 terabit on a single fiber; before leaving GTS in Europe, Mr. Thompson had put such capability in place. In the United States, what had once more commonly been called the Information Superhighway was still capable of handling such capacity easily; and, according to a chart he projected, a desktop or laptop computer could function at between 1 and 3 gigabits per second (see Figure 26). The problem, as the chart illustrated, was that there was not even 1 megabit of connectivity between the two. The DSL connection he himself used on a wireless basis at his home in Rhode Island was capable of only 128 kilobits per second. Expressing skepticism regarding claims that 30 Mbps would be available on DSL, he declared: "The fact of the matter is, that's our choke point."

This came about because since 1983, when access became a concept, companies seeking access had never been able to get it in a way that allowed them to compete.[4] Mr. Thompson recalled that his former company, LCI International, and Bell Atlantic had agreed in 1996 on the existence of "unbundled network elements." Through the telephone companies' leasing out their local plant at reasonable prices, access would be made available to all, and the telcos' embedded plant would not become stranded investment. But this agreement had "lasted about as long as it took the signatures to dry on the page," and nearly a decade

---

[4]Current speeds are determined by the nature of competition as well as by the copper architecture of legacy telecommunications systems.

Current Copper wire based cable modem and DSL
"broadband" technologies choke information transfer

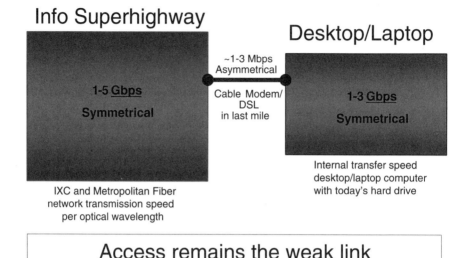

Access remains the weak link

FIGURE 26 Why aren't current services good enough?
SOURCE: Paul Green, FTTH Council Consultant.

later everyone that was involved in that process had retired from the former Bell System, so that there no longer was anyone who understood what the basis of the agreement had been. But, more important, the agreement itself had been "an encroachment on the prerogatives of the incumbents, and therefore access would not be uniform and it would not be ubiquitous." FCC approval ensuring that that didn't happen had been granted about a month before. But juxtaposed to this had been the FCC's decision of the previous week not to subject voice over IP to regulation by the states, which Mr. Thompson equated to "a decision to open up the Internet." AT&T, meanwhile, had announced that it would no longer be a local service provider but that "oh, by the way, they may be in the voice over IP business in the future." While all this amounted to a very interesting process that would be worth following, access was, he emphasized, "the weak link."

Current Capabilities Nowhere Close to Need

Mr. Thompson then projected a chart that placed bandwidth requirements for a variety of residential broadband applications, both business and consumer, against the level of access generally available (see Figure 27). Current capabili-

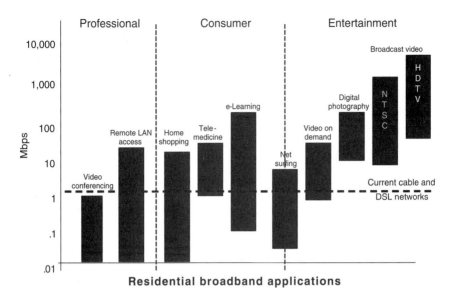

FIGURE 27 Another view of restricted uses: Residential applications/speed matrix. SOURCE: Kim Maxwell, *Residential Broadband: An Insider's Guide to the Battle for the Last Mile*, Hoboken, NJ: John Wiley & Sons, 1998.

ties, including DSL networks, did "not come anywhere near" being able to provide the bandwidth necessary for broadcast-quality television, let alone high-definition TV or even HDTV "on some compression basis," taking account a demand level of three to four sets per home. To make a case for "Big Broadband" or "Ultra-High Broadband," he displayed a chart comparing the speeds of dial-up, DSL/cable-modem, VDSL, 100-megabit, and gigabit services (see Figure 28). The last two, he said, were "what we really need."

Another chart schematized the industry model then current (see Figure 29). Telephone companies had networks, while cable companies had head-ends with capacity to connect to both residences and schools. There had been additions to these capabilities: DSL in the case of the former, set-top boxes in the case of the latter. "But what we still have," Mr. Thompson observed, "is the notion that you've got to have your own network to compete." Even though one often heard it acknowledged, including at the day's symposium, that "the customer wants to be able to buy on a reasonable basis what they're looking for," that need was beyond the current industry's ability to fulfill. He called, therefore, for change: "It's time that we recognized, as the Irish have, that we should be in public-private partnerships to provide access networks." For access networks were not a "nice-to-have" but a "required" facility, instrumental to towns, states, and the nation in "creating the kind of environment that they want to have for economic

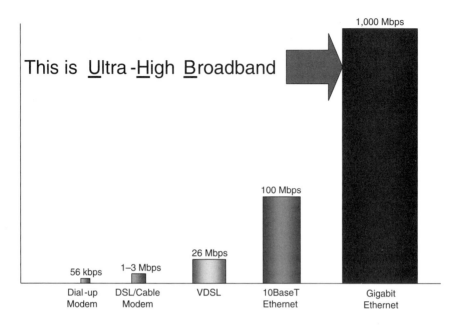

FIGURE 28 Beyond the interim solution: What is the solution for "big" broadband?

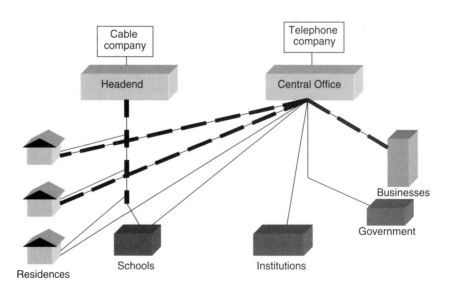

FIGURE 29 Current industry model: If you want to compete, bring your own network.

development to take place." In 1986 he had called in a speech for a "second divestiture," one that would allow the access network component to be separated from the services being provided. "Now we're at the place where, economically, it is viable and, technologically, all the pieces are in place."

Fiber Network as Public Utility

It was for this reason that around two years before, confident of having provided for his children's future and eager to "give something back," Mr. Thompson had formed iTown. The company was "very rapidly" pursuing a strategy that began with establishing "not a central office serving a small community but a central facility serving as a hub for a community of economic interests." The hub would then provide access to a non-profit public-private partnership that would serve as a utility, lighting the fiber but not providing any service on that fiber except those municipal services that the town or community chose to provide. The network would be open to any and all service providers with an IP basis—be they telephone companies, cable companies, a Microsoft, or somebody wanting to provide films or games. Communities should choose to build networks just as, a century before, they had chosen to build roads, just as they chose to build sewer systems or any other element of essential infrastructure.

iTown could play its role of providing the network when there was no one else to do so in towns all over the country that were being either underserved or, in many cases, not served at all. The United States should change, Mr. Thompson argued, such that it would bring existing technology to people rather than forcing them to move in order to have access to it (see Figure 30). The network he

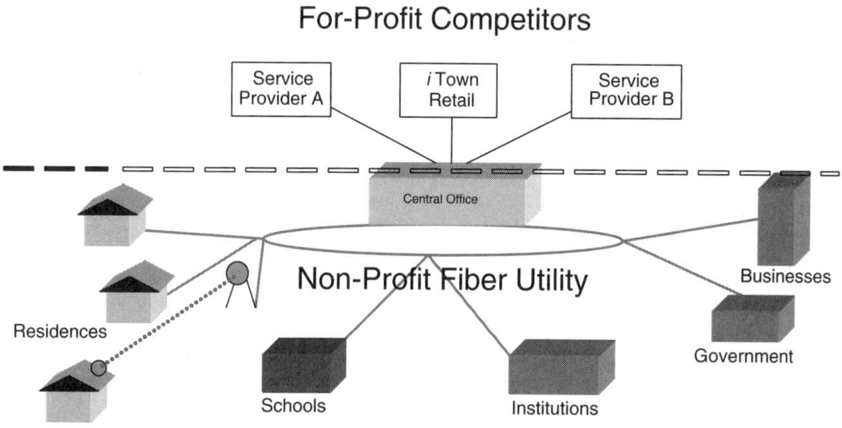

FIGURE 30 Open network architecture allows multiple service providers: All service providers use network under same prices and terms.

described would not only be open to all service providers, it would be used by all under the same terms and at the same prices. The community would control the network assets through a non-profit, to which iTown would provide professional management services.

Access Charges Approach Service Charges

iTown's creation was a response to the fact that "access charges, quote unquote," incurred by customers in some cases far exceeded the service charges themselves. The national average for what was known in the industry as a "triple play"—the purchase of Internet, telephone, and cable-TV services from the same provider—was about $139 per month (see Figure 31). Of that, said Mr. Thompson, $65 per month was billed for "access" to a network "that doesn't cost anywhere near that amount" unless not only past amortization programs but also "all the capital that's been invested in facilities that are no longer valid" are figured into the equation. Displaying figures reflecting current average investment costs associated with building networks in communities of 62,250 and 27,500 inhabitants that provided a fiber link to every home, he noted that payback for the basic cost was projected within 10 to 13 years (see Figure 32); these projections did not include the use of broadband wireless, which he expected would also enter the mix in the future. He challenged the notion that such a fiber network would be obsolete within 10 years, saying the networks iTown planned to build would have "gigabit and multiple-gigabit capabilities." The model, he said, was economically viable and would offer a small town the ability, acting as a local utility, to provide access with "today's and tomorrow's technology."

iTown was making headway in its endeavors, which, however, were not new: 90-odd communities across the country had gone forward with one form or

Access and Service Provider Division of Revenue			
Service	Avg. Total Cost ($)	Access ($)	Service Provider ($)
Local w/Features	36.00*	15.00	21.00
Long Distance	12.00**		12.00
Total telephone	48.00	15.00	33.00
Cable	46.50**	25.00	21.50
High Speed Internet	44.22***	25.00	19.22
"Triple Play"	138.72	65.00	73.72
% of Total	100	47	53

FIGURE 31 Where does the revenue come from?
SOURCES: *FCC data; **Estimated from FCC data; ***ARS, Inc.

Community Size (Population)		62,250	27,500
Required investment		$35M	$20M
Cost per FTTP Distribution and NID	Passed	$997	$1,162
	Served	$1,995	$2,325
Dedicated access revenue (10 years)		$121M	$58.4M
Cash Flow (10 years)		$54.7M	$20M
Take rate and Financial performance	Breakeven	28-30%	~35%
	self-funding	38-40%	38-40%
	redirected income	50+%	50+%
Years to debt payback		10-12	11-13

FIGURE 32 FTTP networks in small communities financially viable.

another of fiber projects. As an example he offered Utopia, an effort to build networks in a number of Utah communities, among which the city of Provo was beginning to move ahead with an active network. Only the previous week, Mr. Thompson's group had announced Opportunity Iowa, a new undertaking involving some 83 towns in the state that had gotten together to provide municipal networks; iTown was working on similar projects in other states as well. "The whole truth here is that we as a nation have got to move," he stated. "We don't have the time to take the next 20 years to get one-third of the households in the country on broadband capability." The consequence of delay would be the continuing erosion of the country's technological base and its ability to maintain its leadership.

THE WIRELESS WILDCARD

David Lippke
HighSpeed America

Offering an overview of his talk, Mr. Lippke said he would identify within the domain of wireless what he called "real wireless broadband" and state what its current status was, what challenges it faced, and where it was going. He might offer some projections for its future as well.

Elusive Definition of 'Wireless'

The term "wireless" was very confusing. Did it refer to a cell phone? to a phone in one's house? to radio frequency identification (RFID)? "What kind of

broadband are we talking about?" was the question—"if we are talking about
broadband at all." For the purposes of his presentation, Mr. Lippke specified,
"wireless" would refer to broadband wireless access (BWA), fixed wireless access
(FWA), wireless ISPs (WISPS) large and small, and metro area networks. Left
out of consideration would be satellite access, indoor wireless such as WiFi,
cellular, or such technologies as 2G or 3G. The capabilities and speeds of these
latter wireless venues could be expected to converge eventually—according to
projections he had recently read, perhaps in around 25 years—but, for the
moment, he would limit his remarks to the former group of technologies.

Broadband, as Mr. Lippke was defining it, was characterized by access speeds
of anywhere from 250 kilobits per second to 20 megabits per second, with the
average probably running at round 1 Mbps. The 250-kbps rate could generally be
attributed to artificial constraints, he said, explaining: "People can turn the bandwidth limits down and still get the money." Also characteristic was low latency,
not of the fiber type but anywhere from single-digit to a few tens of milliseconds.
Where such wireless broadband was available (and particularly in rural settings)
it supplanted satellite, as people would always convert to it from satellite because
its broadband quality was much higher than satellite's. It was being implemented
using WiFi; although WiFi was not designed for outdoor use, it was sometimes
being applied outdoors, pushed up to 10 miles with directional antennas or proprietary protocols. The hubs, commonly pretty small, were often collocated on
businesses or even residences; the cell radius would run anywhere from 1 mile to
5 miles with exceptions at both ends. To illustrate, Mr. Lippke showed photo-
graphs of the installations that brought access to his own home. A flat-panel,
high-gain antenna that was mounted on a post near his house communicated with
antennas on a mountain ridge a couple of miles off, providing around 1 megabit
of connectivity with more available if it were selected.

Wireless Broadband Suppliers in Flux

Although admitting it was difficult to ascertain the true numbers, Mr. Lippke
placed wireless broadband operators in the United States at 2,000 and called an
estimate of between 500,000 and 1 million subscribers "safe." One new operator
was reputed to be "popping up" per day nationwide, but "what you don't know is
how many are dying each day," he said. Such deployments were not, on average,
very large, even if this was starting to change with the arrival of bigger players in
the market. While some operators, particularly in the Midwest and other regions
with conducive terrain, boasted several thousand subscribers, others were mired
at the level of 100–300 and having a hard time reaching critical mass.

To date, wireless broadband had been successful in the T-1 and fractional
DS3 replacement businesses; in both rural and urban areas, it had been able to
compete very strongly on both cost and speed of delivery. The general goal for
these providers was to have, say, 10-Mbps service turned up within 72 hours of a

contract's being signed. Since this stood in stark contrast, both in timing and cost, to service "that one might get from the phone company," Mr. Lippke said, "this is a definite sweet spot for wireless." It had often been very successful as well in uncompeted rural markets, and even in competed small-town markets with particularly conducive geography.

Spelling out the Challenges

Current operators were facing a number of challenges:

- **Relatively high per-subscriber costs:** Deployment, estimated at anywhere from $300 to $1,000 once hub costs were considered, was bumped up by the high cost of installation, which generally required sending a truck to the site.
- **Technology churn.**
- **Proprietary protocols:** Many of these were being used, as was WiFi, which was not intended for this application.
- **Obstacle penetration limitations:** The majority of frequencies designated for this space—and particularly for the unlicensed space, which was dominant—experienced real problems with obstacle penetration. "Trees and buildings are definitely enemies," Mr. Lippke commented.
- **Accurate coverage forecasting:** The latter problems led to difficulty predicting where there would and would not be coverage, which in turn could create inefficiencies. Multiple visits to a site, not only for installation but also for survey purposes, might be necessary; and there could be marketing backlash from customers who had ordered the service only to discover that coverage was inadequate.
- **Insufficient spectrum:** This had been a particular problem at the lower end of the scale, although there had recently been a fair amount of progress in the 900-MHz space, allowing penetration through trees and buildings. "It's one thing when you're talking to satellites and looking 30 degrees above the horizon," Mr. Lippke noted, "but it's another thing entirely when you're sitting on a rooftop and trying to make it to a tower three miles away that is only 100 or 200 feet tall. Then, there are all sorts of obstacles to be considered."
- **General scale issues:** Because of these operators' small size and the concomitant lack of equipment-buying volume, it was hard if not impossible for them to provide a residential broadband product costing on the order of $25 or $30 per month.

While such challenges remained, what had led Mr. Lippke to dub wireless a "wildcard" was the good news that recent progress had been considerable. "And in 2005 and 2006," he predicted, "we're going to see potentially a real flash-over building up." Important to remember was that Moore's Law applied to wireless no less than to other forms of telecommunications. An article in the *IEEE*

Spectrum of July 2004 had heralded a new law, "Edholm's Law of Bandwidth," holding that wireless data rates were increasing at the same pace as telecom rates in general, although displaced somewhat in time. Wireless data rates would thus reach all the points through which data rates for traditional telecom had passed, it was "just a matter of what applications [were] going to be appropriate at what times." Significant progress had also been shown by: software-defined radios; new antenna technologies, such as MIMO, that had the effect of increasing distances and speeds; and, after many failed mesh attempts, there appeared to be some practical mesh occurring. Moreover, a key need of mesh topologies, density of initial deployment, was being affected by the fact that penetration in residential markets was approaching 50 percent. "When the efforts originally failed," Mr. Lippke recalled, "you didn't have near that density of potential customers." And, as had been mentioned by previous speakers, efforts to increase the amount of spectrum, or to expand the rules for accessing existing spectrum, had also been meeting with success.

WiMAX Seen Having Strong Impact

But probably the greatest impact in this space would derive from the fact that the WiMAX standard had arrived, and specifically the form for fixed and nomadic use known as 802.16-2004. WiMAX had resolved a number of problems with existing wireless protocols, particularly WiFi. There was initial talk of 75-Mbps speeds going to 250 Mbps, and, just in the previous days, articles that mentioned the prospect of reaching gigabit speeds had begun circulating. Other issues occupying the wireless world in the past had been quality-of-service (QOS); jitter on the latency; and graceful loading characteristics (as the number of subscribers in a confined space increased, the network's performance gracefully degraded). Moreover, WiMAX had support for a variety of new approaches to antennas and signal processing.

Even more important, WiMAX itself was a certification mechanism for the standard of 802.16, which would bring down deployment costs tremendously. Since installation would no longer require truck rolls, its cost would approach the norm for indoor installation—dipping ultimately below $200, with self-installation by customers possible—so that deployment over a wide area would become very practical. WiMAX could also help compensate for the fact that "wireless" is an ambiguous term by providing consumer marketers with a brand or label that could act as a point of focus around which understanding could be built. Mr. Lippke remarked that the number of articles on WiMAX had recently shot through the roof; he had seen a tally of 400-plus articles in the second quarter of 2004, a threefold increase from the quarter before. "WiMAX truly is a silver bullet that can lead to sort of surprise attack with wide-area wireless broadband," he said, adding that he rarely gives standards such accolades.

Wireless-Fiber Hybridization Gaining Presence

Speculating on "likely futures," Mr. Lippke said that WiMAX-driven services could come to dominate the space in small towns that was being dealt with either marginally—by small cable operators—or not at all. Service providers would be able to deliver at speeds as good as or better than cable and DSL, he predicted. Observing that hybridization of wireless and fiber to the curb was already under way and that broadband over power line (BPL) was starting to get some traction, he claimed to have seen numerous BPL-based business plans that would involve getting "relatively close to the consumer, then bridging around a number of problems once in close by going wireless from there." Cable and telephone companies would hybridize as well, but this was to be expected in BPL in particular. While contending that there was "a lot more hype than reality" to news of the creation of citywide WiFi networks, which were having to cope with hit-and-miss and scaling problems, he said that WiMAX unlike WiFi was a practical protocol for larger-scale deployment and would make such networks a reality. Although the question of how such antenna and switching technologies would develop was unresolved, Mr. Lippke concluded that, owing to cost changes, wireless might well become a third tier that would compete with cable and DSL in tier-one and tier-two areas.

U.S. CABLE: BRINGING THE BITS HOME

Mike LaJoie
Time Warner Cable

Thanking Dr. Raduchel for inviting him to speak at the symposium, and remarking on how different the opinions had been on the complex topic under discussion, Mr. LaJoie said he wished to talk about the implications of convergence: what it was doing to the industries represented, what benefits it had for consumers, how separate industries might want to address it, and how it might be considered from a regulatory perspective.

'Real' Convergence: Not Devices but Content

The "real" convergence, Mr. LaJoie asserted, was not that taking place between the television and the PC or between the cell phone and the television. It was, rather the convergence of data, voice, video, wireless, public networks, and private networks in an end-to-end infrastructure that was increasing competition across all the industries. The economic rewards that arose from this competition would be what drove continued innovation, the advent of new services, and increased broadband connectivity. Consumers would benefit from this competition, but only if economic models continued to drive the investment. A related point that he deemed key as the networks expanded, connectivity increased, and the availability of broadband became more and more the rule of the day, was digital rights management and the protection of copyright. For it was important to ensure that there was interesting content to disseminate via these networks and that the industries creating it survived as well.

To illustrate his notion of convergence, Mr. LaJoie pointed to the way business had evolved at Time Warner Cable (TWC) over the previous decade. In 1994 the company had offered only one product, multi-channel video. Five years later it was beginning to grow in digital video subscribers and had just started rolling out residential high-speed data service. By 2004 the company's revenue mix had changed through growth in digital video and high-speed data, and it was adding digital phone and commercial high-speed data as well.

Formerly Discrete Businesses Overlapping

Some would call the convergence of these industries the transformation of old lines of business: Where there was once "a big separation" between what the telecom and cable industries did, with satellite a new entrant into the competition, "now everybody is in everybody else's business." Mr. LaJoie posted the following list, saying it was "just the beginning" of the businesses that were destined to overlap, offering "similar kinds of products":

- Cable TV
- Internet
- Private IP
- Cellular
- IXC, CLEC, ILEC
- WiFi
- Satellite
- Consumer Electronics

Competition was increasing, something he viewed in a positive light. While recent announcements that Verizon and SBC planned to build fiber to the home and compete with cable in television services might give some people at TWC pause, he said, they represented "a good thing for all of us because it's going to spur initial investment in the economy and drive new uses."

Cable Industry's Massive Broadband Investment

Focusing on cable, he noted that the U.S. industry had invested about $85 billion in broadband infrastructure since 1996 "incremental to just business-as-usual capital investment." TWC alone had, on that basis, invested $14 billion during the period. This eight-year figure paled, he admitted, in comparison to Verizon's investment, which came close to matching it on an annual basis; but in the latter's case it represented the entire capital budget, while for TWC it corresponded to spending on broadband alone. This investment had enabled Time Warner Cable to deploy with unprecedented rapidity a variety of enhanced consumer services, including high-speed data (HSD) and high-definition television (HDTV), launched in 1996; video on demand, launched in 2000; digital video recorders (DVR), launched in 2002; and digital telephone, launched earlier in 2004.

As a result, the company had added almost 4 million HSD subscribers by the third quarter of 2004 to pass between 18 million and 20 million homes, thereby eclipsing 20 percent penetration. By that time, it also had 389,000 subscribers to HDTV, 1.4 million to subscription video on demand (SVOD), and 709,000 to DVR. "The investment in this infrastructure pays off for our customers because they get a wider choice of products to enjoy," said Mr. LaJoie, and it "pays off for us, as long as the climate continues to be as effective for us to invest in as it has been." In 2004 alone TWC had launched digital phone service in 30 out of 31 divisions.

Displaying a timeline showing deployments of new services in the cable industry (see Figure 33), he pointed out that from 1948 to 1972, when pay-TV came in, all it had to offer was video in the form of community-access television. Its next big technological achievement, addressable set-top boxes, dated to 1990. "Not much happened" for the next six years, but then the industry deployed hybrid fiber/co-ax networks and started offering quite a number of new products, with "many, many more being designed and developed" all the time.

Welcoming RBOCs to the TV Market

The announced entry of SBC and Verizon into the TV market, he reiterated, was "great" for the communications industry in general and for the multi-channel

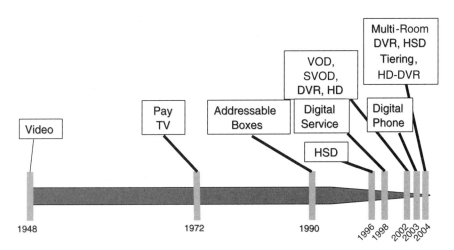

FIGURE 33 New service deployments.

video industry in particular. "More outlets for the delivery of these products to consumers will bring nothing but more investment and a richer set of services," he explained, because "investment in network infrastructure is the key thing that enables all of this." Contesting the opinion of previous speakers, he said the pressure would be on to open these networks to all comers, but he added: "Performance needs to be maintained." No model for additional revenue was in existence "for those who want[ed] to hitch a ride over existing network infrastructure," he said, alluding to Mr. Thompson's iTown venture. Therefore, while the idea of community-owned networks was "interesting," it would be "very difficult to continue making the investment from that posture." Meanwhile, new regulation would need to take into account both the convergence and the amount of investment taking place in these industries, as well as to "provide for economic incentives that [would] ensure continued innovation." As the current system "seem[ed] to be working pretty well," change should be made only after careful deliberation.

Calling the management and control of private networks "critical" to maintaining the quality of service, Mr. LaJoie underlined that it was nonetheless important that private networks and public networks have robust methods for interconnectivity and interoperability. This was being addressed, albeit slowly, in standards bodies, and it might be an area in which regulatory application "could actually help." Integration into an interoperating communications infrastructure would enhance the power of each of these networks. "The more devices connected to the network, the more valuable it is," he explained, and "the more inter-connected networks there are, the more international or national value there is in that infrastructure." Over time, bandwidth and connections—the industry's primary drivers—would tend to become increasingly commoditized. As a consequence, distributors would have to be able to add content, commerce, and other rich communications elements to their offerings.

Consumers' Mobility Steadily Rising

Wireless was having an undeniable impact, Mr. LaJoie said, citing the trend, made so obvious by subscribers, toward consuming content, communications, and information on the move. For network owners, this might present the main challenge of the future: While customers were still willing to consume products at home, more and more wanted the ability to consume them on the move as well. "Nomadicity" and mobility would become increasingly important components of communications infrastructure in the future, requiring far more investment than had the hard-line broadband infrastructure of the last decade.

Concluding, Mr. LaJoie raised the issue of digital rights management. Even though TWC's business, delivering content, did not involve it in owning copyright, he expressed his concern for maintaining the integrity and ownership of copyright. An economic model that safeguarded products and protected their producers' rights to financial return was essential to the continued development of interesting content. A degradation of the notion of copyright ownership had accompanied the advent of digital distribution, he stated, as making high-quality copies and redistributing them had become exceedingly easy. This was another critical area, and one that was "certainly appropriate for regulatory attention."

DISCUSSION

Dr. Flamm opened the discussion by noting that the panel had provided "a rare example" of unanimity in agreeing that broadband was important to the country's economy, its society, and, to some extent, its national security. There was likewise unanimous agreement that the United States had fallen behind other countries and was not necessarily catching up. Substantial variation was to be seen, however, in the panel members' prescriptions for dealing with the nation's lag. While characterizing one of the solutions presented as complete deregulation of the local loop, he pointed out that a complete re-regulation of the local loop had also been put on the table, and that there had been some advocacy of "'relax, don't worry, be happy' as one of the potential policy paths as well."

Claiming the right to the ask the first question as his "moderator's prerogative," Dr. Flamm began by recalling that more than one panelist had spoken of observing a tendency to dismantle some of the opening up of the local loop—for instance, the unbundled network elements that had been a centerpiece of the 1996 Telecommunications Reform Act. He then solicited all the panelists' expectations for future developments: Was unbundling network elements as the vehicle for opening up the Net "essentially dead—and just put flowers on its grave, and move on to something else—or [was it] something that should be resuscitated?" A second question, inspired by the discussion of municipally owned fiber, was whether an increase in regulation at the state level, including the passage of legislation, would limit such initiatives. Finally, he asked the panelists to reflect on the innovative forms being taken by the regulation seen "popping up" at the state level.

Unbundled Network Elements Already Passé

Mr. Thompson, speaking as the author of a portion of the 1996 act, said he "would be the first at this point to put flowers on the grave of unbundled network elements." He said the pertinent provisions of the act had been used as a "lightning rod" to draw a large number of CLECs into the business, based on the notion that "somehow there would be regulatory oversight and a pricing structure." Most of the incumbent operators would call it robbery and argue that it should never have happened, he stated, even as he reminded the audience that these incumbents were the very ones who had agreed to the process in the beginning as a way of getting the act passed. The unfortunate reality was that the regulatory environment in Washington had been greatly affected over the years by the power of the incumbents on all sides; it was therefore no surprise that the notion that open access would rest on a basic, agreed pricing structure had been dead on arrival.

Open Access from the Incumbent's Standpoint

Expressing his disappointment at the turn events had taken, he explained the case from the incumbent's point of view. As the local loop became commoditized, as voice communications increasingly dropped in price, and as such competitive sources as satellite and cable came into the market, it was becoming clear that the incumbent telephone company would be left with plant that was no longer producing revenue. Stranded plant was a fundamental issue; it had been one of the great fears after the breakup of AT&T. And the issue for the incumbent then became how it could shift to getting at least some revenue from that stranded plant; for that would allow it, under its economic model, to invest in future capacity, whether it was going to sell the output on a wholesale or a retail basis. Admitting to having been "on the opposite side" as chairman of Ireland's telephone company, Mr. Thompson said he had wanted to encourage measures ensuring that such facilities "would always be used and we could be the provider at the local level of those facilities." As circumstances evolved, however, regulation had to address the incumbent's obligation to become increasingly open. "If you were going to treat it as a local utility," he declared, "then it should be regulated as a local utility." This, in essence, was what he was proposing.

Mr. Wegleitner, while cautioning that "throwing a technologist into a debate on unbundling [was] probably a very dangerous thing," permitted himself the following observation: The general objection he had heard on the part of the RBOCs, rather than being to the idea of wholesaling access as part of unbundled network elements (UNEs), was to the setting of prices that were "not compensable for the facility." The crux of the matter, from his point of view, not so much the policy of unbundling access as the amount of revenue that could be obtained from selling it as a wholesale facility. Second, the point of the process, as he understood it, was to encourage facilities-based bypass: The facilities—in the case under discussion, copper loop—had been made available so that CLECs might establish some traction in the market, but the ultimate goal had been to spawn competition at the facilities level. This, he stated, "really did not materialize." While there had been continued effort to work the use of the unbundled element as a business, in the end many such business models had simply failed.

'New Wires, New Rules'

Mr. Wegleitner then shifted perspective from the past to the future. Some of what had been tried with the copper network had worked, some of it had not. Saying he was speaking frankly as a technologist, he described the copper network as already "pretty much obsolete" in any case and said it was "going to go away." But what rules should apply to the new facilities that Verizon, along with other players in the business, were putting into the ground? The future would see a move into broadband facilities, among them broadband to the premises. This raised the question of how new construction—"which, quite frankly, anyone could undertake"—would be regulated. His own company's position: "It's new wires, it should be new rules."

Disconnect Between Network, Service Layers

Mr. LaJoie, suggesting that his listeners draw their own conclusions regarding both the effectiveness of unbundling network elements and the associated regulatory impact, identified network owners' main future concern as a "real disconnect between the services layer and the network layer" rather than unbundling. Companies that provided services to subscribers but owned no network infrastructure other than their server complexes—he named Google, Yahoo!, and MSN as examples—had arrived on the scene, grown, and proved successful. He called their viability "a testament to the fact that there are effective models out there for making sure that new companies, and new communication products and services, can be devised separate from who owns the network." This was "one of the beauties of the public Internet," which promised even richer and higher-bandwidth versions of such companies with further growth and the advent of Internet Protocol Version 6 (IPv6). The issues of unbundling and providing access to the last mile had become a moot point, he said, "because you don't need access to the plant."

He recalled meeting Masayoshi Son, the founder of Yahoo! Broadband in Japan, who had driven a huge penetration of new products and services without owning any physical network himself just by leveraging existing infrastructure. One reason Mr. Son had been able to do that was that he had pinpointed a niche in Japan's regulatory arena, whose climate is "completely different" from that of the United States. Mr. LaJoie acknowledged that Mr. Son was "now a billion dollars in debt and losing more every day," but, calling him "an amazing guy," he predicted: "He's going to turn that around." In this country, he reiterated, unbundled access to the physical plant was going to be far less significant than the separation of the network layer and the services layer.

Competition or Consolidation?

Dr. Ferguson said that those who knew his opinions on the subjects under discussion would experience no surprise at his disagreement with the previous two speakers. Noting that his background in the competitive information-technology sector had shaped his temperament, he said that he shared the overall goal of a deregulated environment. "But that environment can and should be deregulated," he stipulated, "only once it becomes open and competitive, which it currently is not." While it was "very much in the incumbents' interest to portray themselves and each other as [being] as competitive as possible," he observed, "if you in fact look at the structure of this industry over the last ten years, it has been getting steadily less competitive." He issued a warning: "If you deregulate a monopoly, you get a deregulated monopoly."

Detailing the "wide variety of ways" in which he believed the telecommunications industry to have been concentrating, Dr. Ferguson began with Internet access. Under the dial-up regime there were thousands of ISPs, but "when everybody converts to broadband, there are going to be two ISPs." Asserting that the cable industry had been consolidating rapidly, he charged that it had become "an oligopoly of about half a dozen diversified, vertically integrated industrial complexes." All of these owned proprietary content assets; alluding to Mr. LaJoie's comment that Time Warner Cable's business was delivering content and that it was thus not involved in owning copyright, Dr. Ferguson declared that TWC's corporate parent was "very definitely concerned with copyright" as owner of Warner Communications, of Warner Bros. Studios, and, through Time Inc., of text-based magazines.

Dr. Ferguson traced a similar consolidation among ILECs, saying their number had dropped from somewhere between seven and nine to four. Commenting on the UNE regime, he moved to cast doubt on the ILECs' complaint that they were being subjected to unfair pricing, with the prices of unbundled network elements set unfairly low. "If that [were] true," he argued, "then the most logical thing in the world [would have been] for each of them to purchase those incredibly cheap unbundled network elements from each other, integrate into each others' geographies, and compete with each other." That none of them ever did that, he stated, meant that "a certain degree of skepticism" was in order when viewing "those claims, and the structure of this industry, and the kind of conduct it has engendered."

Concluding, Dr. Ferguson cited several issues—"the precise way in which an open-architecture industry can and should be achieved, exactly which interfaces should be open, whether there should be divestitures or not"—as meriting debate. Noting the potential to choose among "a wide array of possibilities," he nonetheless emphasized his misgivings regarding the status quo and the direction in which the industry had been moving.

Dr. Flamm, responding to Mr. Thompson's request to comment, also asked him if he would discuss whether municipal fiber networks had not already been preempted by state law in a number of places.

Regulatory Underpinning to Access Cost Structure

Mr. Thompson said he would begin by making a point that, though "critically important," might escape some who were younger than he: In 1983, when the access regimen was put into place and the first policy debates occurred, "there was no such thing as the Internet." The TCP/IP protocol did exist, and there were four companies using it; one of these, Western Union, had 750,000 employees who were members of the Communications Workers union. During discussions of the use of the embedded network and of who should pay for it, there was a "very great hue and cry on Capitol Hill" that came back to the FCC, which established "enhanced services." These services—the use of the TCP/IP protocol and of telex on the backbone, as it was being employed at the time for TWX and other functions—carried no fees for access to the public switched network. And the Internet service providers of ten years later, when the Net was taking shape, still paid nothing for access despite a "huge outpouring of objections from every one of the incumbent telephone companies." In fact, even in 2004 an ISP's access to the public switched network was without cost. "What we have seen as the advance of the Internet in our society," he concluded, had therefore enjoyed a "huge regulatory underpinning."

States, Incumbents Ganging up on Non-profits?

Turning to Dr. Flamm's question regarding state laws' preempting municipal fiber networks, Mr. Thompson asked: "Is it any great surprise that the incumbents—which in many states are very close to the regulators and play a fairly major role in political campaigns—have put forward proposed legislation in virtually every state making it illegal for a municipality to compete with them on the grounds that the municipality would come to the business from a non-profit and governmental point of view?" He stressed that his company's was not a competitive offering but a significantly different package, based on the premise that the municipality had a right to provide an access network—as it does in the case of roads, sewers, or water—but would not provide services unnecessarily in competition with incumbent franchises.

Each state's laws were different, Mr. Thompson explained, in line with whether the state had been nurtured or neglected by its incumbent providers of services. Those towns participating Opportunity Iowa, whose advisory board included two former governors and numerous university presidents, felt "the time had come to make a point," he said: "They have been overlooked by their incumbent carriers and their incumbent cable operators to the point where they are losing their brains, the graduates of their schools, to other states." In 2003, a bill had been put before the Iowa legislature by the largest incumbent cable provider and the incumbent telephone service provider that would have made it illegal for any community to own a telecommunications network. The same had been done in other states, including Wisconsin and Kansas; Opportunity Iowa had been crafted as a non-profit, political effort in order to deal with such pressures— which were being seen all across the country—before the legislature. The following year or two, he predicted, would see "a very interesting debate" about the role of the community in providing essential facilities for their residents, for, he stressed, network access is an essential facility.

Seeds of a Consumer Rebellion?

Dr. Flamm opened the floor to questions, and John Gardinier, who identified himself as retired, commented that he was leaving broadband because of the monopolies. "I know some consumers are leaving cable completely," he said, "and a certain number are leaving their phone providers to go strictly cellular." He asked Dr. Ferguson to state his reaction to Mr. Thompson's endeavor—speculating that he would see it as a step in the right direction, albeit one with shortcomings—and to reflect on the idea that unless the industry looked to a different model, it might face a consumer rebellion.

Dr. Ferguson, characterizing as interesting the question of whether and at what point the broadband problem will become politically salient, identified two possible sources of rebellion: consumers and the technology sector. The last mile already had become a significant drag or drain on the growth rate of much of the technology sector, he said, pointing specifically to personal computers. "In private, a number of those companies and the people who work in them will tell you they're rather upset about it," he recounted. "But, unfortunately, the telephone companies are frequently their largest customers—sometimes the cable companies are as well—and both are politically powerful, so they have to be careful about how they proceed." He pronounced himself as "not terribly optimistic" that this would become a political issue soon.

Addressing the status of municipal networks, Dr. Ferguson said he would concur with Mr. Gardinier's formulation: that they were a good thing but were unlikely under prevailing regulatory, political, and economic conditions to make up for the system's other problems. Tax incentives for the construction of municipal networks could be useful in a reformulation or improvement of broadband policy. There would, however, be a requirement for continuous technological improvement of those systems over a long period of time, and in general municipalities were not considered to be the best stewards or managers. He did not, therefore, believe that municipal networks would replace other portions of the system.

Prospects for a New Telecommunications Act

Mike Nelson of IBM asked the panelists to reflect on a question raised by Mr. Tenhula earlier in the day: whether there would be a new telecommunications act. What might actually drive a rewrite of the 1996 Telecommunications Act? Was it possible that giving this serious consideration would only generate more uncertainty in the marketplace—considering that, as the details were hashed out and after they were finalized, it would take the courts another five years to figure out what those details meant?

Mr. Thompson, saying he had spent some 15 years on the 1996 Telecommunications Act beginning with what was called the Bell Bill in the late 1970s, offered to take off his shirt and display his scars. A telecommunications bill ranked only one step below a trade bill as measured in money spent, both by proponents and by opponents. Any would-be competitors to the incumbents were, at the moment, weak enough financially and sufficiently dispersed that there was no strong need to propose a bill. The exception was that a bill might come from the telephone or the cable companies, but he saw that as unlikely because most FCC decisions on the issues attendant to the 1996 Act had, to date, been in favor of the incumbents.

Mr. Wegleitner, while demurring on the question of whether another tele-communications act might be coming, stated that both technology and "the way the industry has shaped up" had outrun the 1996 Act. He said that something needed to be done but expressed uncertainty as to whether another act would be needed to do it.

Factors Behind U.S. Broadband Lag

Turkan Gardenier of Pragmatica Corp. offered a pair of observations regarding the United States' No. 13 ranking in broadband penetration. First, whereas outside the United States the caller paid for a cell phone call, in this country the recipient was responsible for the charges incurred by the caller. Because these charges could add up, she said, she personally did not give her cell phone number out to many people. Second, where a larger percentage of the population lived in apartment houses, such as in the Far East, whole buildings could be wired for broadband, avoiding installation fees and monthly charges for each individual dwelling. Could these factors be contributing to the less-than-optimal use of broadband technology in the United States?

Mr. LaJoie disputed the notion that the charts showing the United States in thirteenth place told the full story. For a number of reasons, some related to the issues raised by Ms. Gardinier, the picture was somewhat less stark than the charts made it appear. Important to keep in mind, he said, were differences in regulatory climates, in the history and condition of infrastructures, in how products were used, and in the concentration of homes. The concentration of population in Tokyo and Seoul, for example, was much greater than in all but a few places in the United States, and so broadband penetration would tend to be greater there. A comparison of penetration there with penetration in major U.S. cities would show that there was not such a discrepancy. But the regulatory climate and the age of infrastructure were also significant. Building greenfield was different from putting in facilities when a lot had already been built out. While infrastructure in Asia and Europe was newer than that in the United States, this country was in the process of making investments in both the cable industry and the telecom industry.

No Denying U.S. Broadband Shortcomings

Dr. Ferguson said he disagreed for a number of reasons. First, even if it were true that geography, population density, or other considerations made a difference with regard to broadband penetration, they did not explain growth rates. For growth rates had been "enormously different": less than 40 percent per year in DSL residential connections in the United States vs. about 80 percent in the rest of the world. Second, the countries that were ahead of the United States, and that in many cases were growing more rapidly even though they were already ahead, were not always countries with markedly different geographical or population-density profiles. Not only was Scandinavia ahead of the United States, so was Canada—a nation almost as large as the United States with a population of only 40 million. Third, a comparison of price to performance in urban areas would show bandwidth to be somewhere between two and ten times more expensive in the United States than it was in much of Asia. Arguing that such facts undermined Mr. LaJoie's claim, he insisted that the United States' problems in bandwidth penetration were real.

Connections Between Communications, Transportation

Mr. Hellman said he would discuss the relationship between communications and transportation but he wished to preface his comments by saying that he considered outrageous that many accepted DSL or cable-modem as broadband; instead, he asserted, these were "the sales tax on broadband." Then, speaking as a real estate developer, he noted that because the long-term fixed assets he constructed—buildings—do not move, wireless was of interest to him mainly as a subset of a much bigger picture. Since it could be deployed very quickly, wireless might indeed be a step in the evolution toward a new wired network; but its particular value, mobility, was irrelevant in the case of buildings.

The economic significance of the relationship between transportation and communication, Mr. Hellman declared, was greatly underappreciated. Transportation was extremely expensive, and the expense could arise at many different levels:

- building roads and bridges was expensive;
- building rail was expensive;
- oil and gas were expensive, and they had international political implications as well; and
- environmental compliance could be expensive, and arduous as well—the Washington, D.C., metropolitan region was currently in violation of the federal Clean Air Act.

The way to begin solving such problems was to reduce the amount of time people spent in stop-and-go traffic. But, "if you're going to do that," he said, "you've got to give them an alternative way of getting the job done." He called desktop videoconferencing an absolute necessity in this regard but noted that it required sufficient bandwidth that "you can see a good picture and see what you're talking about at the same time." Bringing together the policy discussions on transportation, which entailed massive expenditures by both the federal government and the states, and telecommunications might be a way of starting to come to grips with these issues, he suggested.

Extending Great Teachers' Reach

Related in Mr. Hellman's mind was another public-interest issue, that of improving education. The decline of urban schools was much discussed, but less talked about was the fact that "the suburbs [were not] going to give up good teachers to the cities." Because truly great teachers were limited in number, "a hybrid of computer and communication networks and broadband and interconnectivity" would be required if the best teachers were to be made available to everybody. He therefore recommended expanding the policy debate.

Mr. Wegleitner stated his agreement with Mr. Hellman on broadband's potential for helping solve the nation's energy problems. While he hesitated to make a detailed statement connecting the construction of a fiber-optic network with the economics and social impact of reducing emissions from gas-powered vehicles, he called "being able to move the business to the house—and, in fact, transparently providing the virtual office—absolutely part of the vision of broadband as we see it." Upstream was one of the key components; downstream was arguably O.K. at a few megabits, but symmetry was needed and 10 megabits or more would be required in order to do the job correctly. It was for this reason, he said, that Verizon had adopted as its platform for fiber deployment the G.983 passive optical network, which "can take a residence and make it look like an office in a remote sense."

Mr. Wegleitner affirmed Mr. Hellman's observations on the importance of distance learning as well. Verizon had built a number of educational networks using the more traditional ATM frame relay technologies. He mentioned Network Virginia, on which Verizon had worked in conjunction with other carriers to tie together a multitude of universities; extending its reach to other schools as well was a primary objective. A similar network was online in New Jersey.

RBOCs' Consolidation: Legal, Economic Perspectives

Mr. Nelson recalled that Dr. Ferguson and Bob Metcalf had a few years before suggested the need for an antitrust suit against the Regional Bell Operating

Companies (RBOCs), since they clearly weren't competing with each other.[5] He asked Dr. Ferguson whether these declarations had prompted calls from:

- lawyers who wanted to explore the possibilities, since such a settlement would have made them far wealthier even than the tobacco lawyers;
- the Justice Department's Antitrust Division; or
- economists who wanted to do a study that would show there actually was a case.

There had indeed been discussions on these questions with the people who ran the Antitrust Division in the Clinton administration, Dr. Ferguson said. "There was clearly some interest and concern," he recalled, "but also an acknowledgement of the rather significant political barriers." To highlight the size of these barriers, he stated that the industry's incumbents "literally spend more money on lobbying and litigation than they do on R&D." In this, they are behind only the energy industry and, possibly, the cigarette industry.

When Mr. LaJoie objected that the cable companies spent far less, Dr. Ferguson acknowledged this and specified that, by "incumbents," he had meant the ILECs. He asserted, however, that spending by the cable industry was "increasing rapidly." Returning to Mr. Nelson's question, he added that he had received calls from private attorneys exploring private antitrust suits and noted that one "semi-serious" attempt had been made but it had, to his knowledge, not gone anywhere. There had been no calls from economists.

Dr. Flamm then thanked the panelists for an excellent discussion.

---

[5]We may note in this regard that a firm in one geographical market is under no obligation to to compete against firms providing similar services in a different geographical market. To the extent that the Regional Bell Operating Companies operate in separate geographical markets, this issue is not relevant.

The Waterfall Effects

INTRODUCTION

Cherry A. Murray
Lucent Technologies

While Dr. Murray explained that the title "Waterfall Effects" indicated that the panel would be devoted to some newer applications in telecommunications, she also reminded the audience that "voice is the killer app, and voice will continue to be the killer app." Still, she noted, in the areas of the world where "broadband is just rampant"—in South Korea and Finland, for example—messaging was becoming increasingly popular, especially among the younger generation.

As the first speaker, she introduced Mike Nelson of IBM.

MOVING COMPUTING TO THE GRID

Michael R. Nelson
International Business Machines

In his talk, Dr. Nelson said, he would cover "what's beyond broadband, why we need to keep continuing up the technology curve to produce faster and faster networks, and what we will do once we get there." To start, he encapsulated his main points in what he called "bumper stickers," easily remembered summaries seven or eight words in length whose value he had learned while working on Capitol Hill and at the White House. His first point so expressed, "It's not just about email and the Web," was intended to signal that the Internet had entered a third phase. This transition was being made possible by grid computing, autonomic computing, pervasive computing, and open standards, all of which he planned to address further.

The initial 20 years of the Internet, Dr. Nelson recalled, were marked by one-to-one applications. Its very first user, who was located in Los Angeles, attempted to log on to a computer at Stanford; because the system crashed before this user could type "log in," the first message on the 'Net read "lo." Such one-to-one messages, whereby a person talked to a computer or to another person, were typical of the Internet's first two decades, constituting most of its traffic until about 1990. Then, the advent of the World Wide Web precipitated a fundamental change: Through the addition of one-to-many communications, the Internet became a "broadcast medium." This important step resulted in a remarkably sharp increase in the amount of Internet traffic; for a short period, it doubled every four or five months, all because of the Web.

Internet Undergoing a Pivotal Transition

The present was a similar moment, said Dr. Nelson, arguing that the Internet was undergoing another pivotal transition to become a "many-to-many medium." Napster, the first example of this phenomenon, had shown "what could happen if you took a million people and hooked them up to a network that tied together 300,000 PCs all operating as a single system." When users went onto the Napster network looking for an obscure Beatles recording, they didn't care which computer actually had the bits that they wanted: "They knew only that somewhere out there on the network would be the answer."

In this way, Napster had demonstrated the power of a new paradigm—which in its own case, unfortunately, had been illegal. But that same principle had begun serving as a base for other innovative technologies. Dr. Nelson praised his employer, IBM, as a leader in one of these, known as "The Grid." This technology allowed not only systems that had music files to be hooked together, but also systems that shared other types of data, software, and—perhaps most important—computing power. Likening the result to the supplying of electricity by a utility, he said that a user logging onto The Grid could obtain access to far more computing power than was available on that user's own systems.

Peer-to-Peer Computing: Promise and Limitations

To illustrate the variations of distributed computing, Dr. Nelson displayed a graph with the number of nodes on a grid plotted on the y-axis and the power of

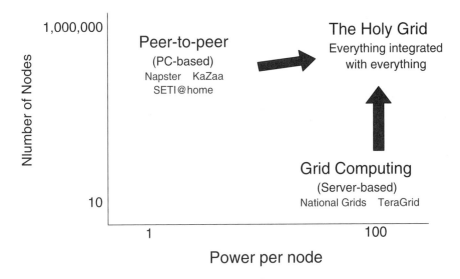

FIGURE 34 Many flavors of distributed computing.

each node on the x-axis (see Figure 34). He first addressed peer-to-peer computing, in which PCs are tied together to provide hundreds of millions of dollars' worth of computing power that runs software aimed at a specific problem. Naming Napster, KaZaa, and SETI@home, he commented that while each handled its task well, it was unable to go beyond that single function to perform others.

He focused on the example of SETI@home, describing it as a screensaver that harvests all cycles on a user's laptop or desktop that are not in use—a considerable bounty, considering that a typical laptop is used only about 2 percent of the time and that most of its power is wasted even when it is in use. "With SETI@home," Dr. Nelson explained, "you get a little piece of radio-antenna data from Puerto Rico, and your computer tries to find some kind of consistent signal in that data to see if we are getting a signal from intelligent life on Mars or in another galaxy." As 500,000 people had downloaded the screensaver, it had generated an amount of computing power that would have cost over $100 million to purchase.

In grid computing, situated opposite peer-to-peer computing on the graph, fewer nodes are tied together. But because of the size of the machines—large servers and storage systems, even supercomputers—at least as much power is generated. In addition, since the systems involved in grid computing are more tightly coupled and more general-purpose, they can do more. Dr. Nelson reserved his greatest excitement for what he called "the next step: the 'Holy Grid,' where everything is connected to everything, running common software, able to tackle a wide range of problems."

The 'Utility Model' of Computing

Positing the notion of computing as a utility, Dr. Nelson discussed his vision of The Grid in light of the history of electrical distribution in the United States. In the early decades of the last century, most American companies had a vice president for electricity, who was in charge of making sure that each factory had working generators to supply the power needed. Once electrical utilities showed that they could provide power more cheaply and more reliably, however, few factories continued running their own generators. With the advent of The Grid, companies large and small would be able to proceed on a pay-as-you-go basis. "They will be able to buy the computing power they need and get the software they need over this grid of network systems," he stated. "It's got everything a normal laptop or server would have: data, applications, storage, processing power."

What will eventuate, Dr. Nelson predicted, is a unified system that will be managed as such and be able to provide services to all who tap into it. Service will be better and efficiency higher as a result. "You make much better use of your systems," he said, "because rather than a laptop or desktop being in use only 5 percent of the time or 3 percent of the time, it can be part of a larger system and contributing excess cycles to the grid." Even a typical corporate server is in use only about 30–50 percent of the time and is thus a potential source of power to be harvested. In addition, because The Grid is to be managed as a single unit that will unify "different sites, each managed by different people running different software," security will increase and complexity diminish. In this "new world," systems and software will be virtualized: The user will be able to log on to the grid, draw data from several different sites, pool it, process it using computing power from several other sites, and then output it somewhere else. This presents a powerful opportunity for collaboration. By allowing all its different sites to tap into the global grid, a company would be giving all employees access to its most powerful tools, something not possible with the current Internet.

The First Steps Toward 'The Grid'

The first step in the development of The Grid has been the creation of intranets by companies that take existing hardware, tie it together with high-speed systems, and use the resulting network as a grid. "They don't have to buy any new servers or storage systems," Dr. Nelson said, because by running "software that ties their systems together they can double or triple the amount of computing power they get out of their existing equipment." IBM tests some of its chips, using what is called the "download grid," whereby employees all around the company back up their laptops and desktops. Although the application may be regarded as mundane, it can be carried out much faster and more economically thanks to the grid.

The second step in The Grid's development is the partner grid, which involves companies tying their systems to those of other companies. The third step is the actual move to the utility computing model, under which third-party grids run by independent companies—possibly IBM, AT&T, or the telecommunications providers—furnish the computing platform upon which thousands of businesses run. IBM had just started some demonstration projects in this area; for one of them, the "Smallpox Grid," about 10,000 IBM employees had downloaded software enabling their computers to do modeling designed to determine whether a particular drug molecule might be used to block replication of the smallpox virus. The project had generated millions of dollars' worth of free computing power for Oxford University, which as a consequence had identified 10 or 12 drugs worthy of further investigation. This software was running on Dr. Nelson's computer as he spoke, trying to match a molecule with the virus to see whether there was a way in which the two locked and, thereby, to identify an anti-smallpox drug that merited testing.

Autonomic Computing and Pervasive Computing

Also part of this vision for future computing is "autonomic computing": systems that are not only self-protecting, self-optimizing, self-configuring, and self-healing, but that also come close to being self-managing. IBM customers, Dr. Nelson said, were experiencing enormous increases in the number of transactions they processed and the amount of data they stored. Unable to hire enough qualified people to run all the systems required, they needed systems that could take care of themselves. "The Grid will facilitate that by making it easier to manage many systems at once," he said.

Another important component of the vision was pervasive computing, something that Dr. Nelson felt had not received sufficient emphasis. It was his working assumption that, five years down the road, he would own literally hundreds of devices and products that interacted in one way or another with the Internet. Many of them would have a radio frequency identification (RFID) tag, others simple sensors; "anything in my house that's worth more than $50 or $100, or that has some moving part, will probably have some way of interacting with the 'Net," he said.

In reference to a logarithmic diagram showing the numbers of computers, appliances, and sensors that have been connected to the Internet since 1990 and projecting them out to 2020 (see Figure 35), Dr. Nelson observed that sensors could be expected to become more numerous than either of the other two within 5–10 years. While many of the world's 1 billion PCs were connected to the Internet, they already lagged cell phones and other devices. "Soon," he said, "we'll have trillions of sensors, and that will be what we really rely on the 'Net for." These sensors will be located all around the world and the data they generate will somehow have to be managed, something he saw as another application for The Grid.

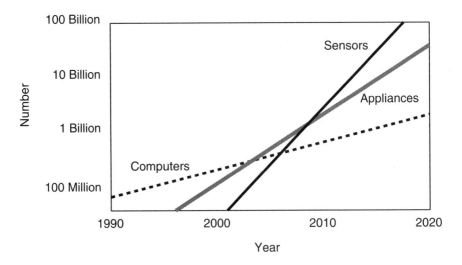

FIGURE 35 Sensors will predominate: Internet-connected devices.

Computing Power Available on Demand

Behind IBM's excitement about The Grid, as well as about autonomic and pervasive computing, is the role they play as building blocks of what the company calls "E-Business on Demand" or "On-Demand Business." The integration of a company's entire IT infrastructure using common standards and common software will make it much easier for the company to obtain the computing power, data, and software it needs when it needs them. Currently, tackling a new problem can take weeks if not months, Dr. Nelson said, because it means ordering numerous servers, having them brought in, having somebody configure them, and getting them up and running. In what he referred to as "this new vision of the future," acquiring the computing power sought will take a "few hours, or even a few minutes—just as, today, if you need some extra electricity, you can just plug something in." The vision, IBM's response to its customers' demands for less complexity, more reliability, and improved security, will require better networks, he acknowledged.

Pulling out another of his "bumper-sticker" phrases, Dr. Nelson estimated the Internet Revolution to be less than 8 percent complete, a figure that nonetheless registered an improvement over the 5 percent of "a couple of years" before. Some 8–10 percent of the world's population was using the 'Net on a regular basis, with the total number of Internet-connected devices at three or four per person in the United States. As many new and exciting applications could be expected to be enabled by The Grid, this figure would rise to "dozens if not hundreds," he predicted, adding: "No matter how you measure it, we're just at the start of this."

Increased Activity Assured, New Policies Needed

Offering a formula to aid comprehension, Dr. Nelson advised those in attendance to "take everything that's already happened—all the new applications, all the new content, all the new money that's been made, all the bankruptcies—and multiply by 12." Realizing the vision of The Grid and the next-generation Internet will require some new technologies and significant investment, he cautioned, as it will entail providing whole neighborhoods with gigabit-per-second networks that are as affordable and reliable as they are ubiquitous. "Getting there is going to require more intelligent, more consistent policies than we have today," he declared, noting that he was far from the first speaker of the day to call for policies that were more consistent. Furthermore, those working toward this vision would "have to look beyond the FCC" if they hoped to address all the issues currently driving decisions, which he summed up with a list he had developed 15 years before and titled "The Ten P's of Cyberpolicy" including pricing, privacy, piracy, pornography, protection (security), policing, procurement, payment, and protectionism.

In fact, they would also have to look beyond policy makers in general. Spending much of his time on standards issues, Dr. Nelson said, had impressed upon him that the next-generation Internet was already being shaped by critical standards that were in development, as well as by choices that the marketplace was making between competing standards. Posting a list of "key technology choices" (see Figure 36), he said that how those issues and perhaps four or five others were decided would not only shape the next generation of the Internet but also determine whether The Grid became a niche application or something upon which almost every company relied on a daily basis.

U.S. Decisions' Worldwide Impact

His final point was that decisions being made on these issues in the United States would have an impact on developments in other countries. It would affect them directly, because the market for new products created in the United States would enable sales elsewhere. It would affect them indirectly as well, because "if we decide to do something stupid here, there are at least 40 countries that will probably emulate our stupidity," said Dr. Nelson, adding: "We have to make sure they learn from our stupidity rather than emulating it."

Furthermore, if The Grid's rollout justifies his expectations, taking the form of a "grid of grids" that ties all countries' information-technology infrastructures together in a global digital economy, future debates about offshoring, allshoring, and allsourcing will make the current one "look pretty tame." In the resulting

- Authentication and directories
- Privacy-enhancing technologies (P3P)
- Digital Rights Management
- Filtering technologies to block spam, porn
- Voice over IP
- Wireless Internet standards
- Web services and Grid computing
- Instant messaging
- IPv6 deployment
- Linking the phone network and the Internet
- Rich media standards (SIP, multicast, etc.)
- End-to-end vs. walled gardens

FIGURE 36 It's not just about laws and regulation: Key technology choices.

environment, any employee anywhere will have the ability to tap into The Grid, and any company will be able to compete with any other using the most powerful tools available. If the Internet led to "the death of distance," then The Grid will mean "the death of geography," because companies everywhere will have access not only to computing power but also to collaborators, databases, new tools, and new software. Opportunity will abound, but so will weighty issues.

Introducing Louis Mamakos of Vonage, whose talk was titled "Is VoIP the Future?" Dr. Murray observed that, already, VoIP was the present.

IS VOIP THE FUTURE?

Louis Mamakos
Vonage

Mr. Mamakos endorsed Dr. Murray's assessment, noting that there had already been quite a bit of uptake of voice over Internet Protocol technology, and said he would be speaking about how the market for this service had developed.

He began an overview of the factors that helped bring VoIP capability to market by observing that the Internet had decoupled the transport of bits from applications. Internet service providers (ISPs) had supplied the pipe to plug the computer into; new, interesting, and varied applications had come from numerous sources. "If we can arrange to have an environment where new and innovative ideas can be tried out," he observed, "interesting results pop out." Recalling "Sturgeon's Law," coined by the science-fiction writer Theodore Sturgeon— "Ninety percent of everything is crap"—he emphasized that, for the remaining 10 percent, that was not necessarily true. Such innovations as the World Wide Web and email had been the product of extensive experimentation rather than of "people going off into a room, thinking really, really hard, and coming up with the answer." This had also been the case with voice over IP, which he described as "something familiar cast in a new light."

Markets, Services Increase with Broadband Penetration

An important enabler had been the growth of broadband. But while broadband is a prerequisite for any such multimedia service, his own company's offering, and voice over IP service in general, are fairly insensitive to the type of technology over which they are run as long as capacity is adequate. The increasing penetration of broadband deployment, globally and in the U.S. (see Figure 37), opened new markets to new kinds of products and services. The reception with which not only Vonage but also other VoIP players had met in the marketplace (see Figure 38), Mr. Mamakos said, indicated that the service's acceptance had moved beyond an early-adopter population to the more mainstream consumer.

Because VoIP exists within a broadband environment, basic assumptions can be altered, including those regarding the way in which the customer interacts with the service. Service provided over the public switched telephone network (PSTN) has in recent years offered such options as call forwarding and call waiting. But, Mr. Mamakos said, changing the provisioning of these features has tended to be a fairly lengthy process: Where it is automated, the interface consists of audio heard in the ear plus a ten-digit keypad on the phone. Voice over IP takes advantage of broadband to present service parameters to customers using a very rich, high-fidelity interface in the form of their Web browser. New and interesting services can thus be delivered that may have been available previously but were simply too unwieldy to control without the customer's having access to a richer interface. "In the voice over IP world," Mr. Mamakos noted, "all these features are just software, and if you look at how voice over IP operators tend to deliver their services, these features are bundled in as a standard part of the service offering." That the customer is not paying extra for touch-tone dialing, caller ID, three-way calling, and other services is, he added, a "side-effect" of the amount of power that VoIP brings to the marketplace.

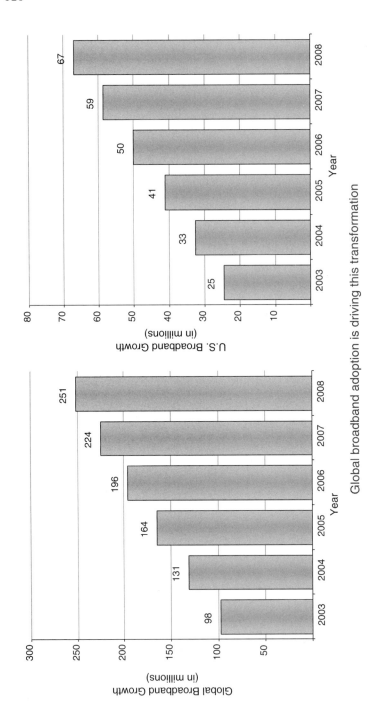

FIGURE 37 Broadband growth.

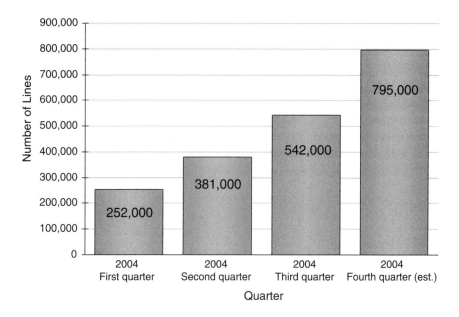

FIGURE 38 Growth of U.S. consumer VoIP lines.

VoIP: Shared Infrastructure, Greater Customer Control

Mr. Mamakos cited two sources of opportunity that arise with VoIP. One is through sharing infrastructure, which comes of chopping up audio into packets and transmitting it over an existing packet-based network. But equally powerful, he contended, are opportunities that come of making the call control of services available on platforms that are easier to program than a telephone switch. This flexibility, in the form of exposing call processing, made it possible for companies like Vonage to try out very interesting ideas, some of which might resonate with customers.

The company would start with features that customers are very familiar with, Mr. Mamakos said, but he suggested that it might then blend familiar elements into novel contexts. As an example, he offered integrating buddy lists from instant-messenger clients with phone service so that customers could control who could call their phone after 9 in the evening. An instance of integrating telephony service with computer capability that Vonage had already developed is "Click to Call," which allows the user to highlight a name in his or her email address book, then click a button that rings both the user's Vonage phone and the person that he or she wanted to call—"thus," he said, "saving that tedious dialing, with its wear and tear on the finger." Conceding that "none of this is really rocket science to figure out," he explained that, nonetheless, "someone has to write the code, and, more important, once you've actually done interesting features like this, you need to be able to expose them in a way that is easy for a customer to manipulate." It is the Web interface that allows Vonage to augment the richness of the services going to market.

Areas of innovation in which Vonage had recently been active included a mobile 911 service akin to that available through a cell phone and E911 capability, which had been the subject of some early trials Rhode Island. Regulatory help in the form of allowing new entrants access to some emergency response infrastructures would be welcomed by providers of such new and interesting applications, Mr. Mamakos said; to date, these infrastructures had been centered around the legacy telephony infrastructure. He posited the development, with the evolution of technology, of a small device outfitted with a panic button that could be used to summon help in an emergency—assuming that such appliances were granted access to the emergency response infrastructure.

Open Standards Credited as Enabler

Mr. Mamakos credited open standards as an enabler of voice over IP technology. Within the Internet Engineering Task Force (IETF) and other standards bodies, fora existed for developing standards to allow many different manufacturers to bring products to market. Especially in the case of the IETF, standardization of the technology had been viewed as more a process than an event, he said, and "typically, standards happen after there's been some demonstration of the technology and production deployments." As part of this ongoing process, interesting ideas are tried out, some of which strike a chord and achieve market acceptance. "Then, the benefit of that early deployment works its way into the standards," he said, "making it easier for other participants to build products and services [that are able] to participate in the market."

VoIP-enabled silicon can be embedded in any device; Mr. Mamakos displayed an image of an object that appeared to be a cell phone but was in reality an appliance integrating a VoIP telephony adapter with a Wi-Fi radio. At a hotspot, this could be used for voice over IP service nomadically, almost as if the user were to carry a dedicated telephony adapter and find an Internet jack to plug it into. Because it is after a mass market, Vonage tries to make the service "as much plug-and-play as possible," he said. And customers are, in fact, simply plugging the "familiar telephony instrument that they are used to having in their homes" into a voice over IP adapter rather than plugging it into the wall to connect to the telephone company. Although they have the benefit of a Web-based interface to control features, for day-to-day calling the interface is the same as always: Users pick up the phone and dial. Last-mile technology is a matter of indifference to most VoIP services, which require only a broadband connection with adequate capacity. Thanks to the interconnects between the Internet and the PSTN, a voice over IP customer can call ILEC customers who have the same legacy phone service that's existed for the last hundred years as well as other Internet-connected VoIP end users.

VoIP Matches Landline Call Quality

According to a survey conducted by *Smart Money*, the quality of a voice over IP conversation was on par with that achieved over a traditional landline phone and typically better than could be achieved on a cell-phone call. Judging this result "not too surprising," Mr. Mamakos pointed out that a typical CDMA cell phone uses "about a 13kb-per-second codec, and you're trying to squeeze as much capacity onto fixed spectrum allocations" as possible, while a VoIP adapter with total-quality voice uses the same 64kb-per-second transport codec used by the PSTN.

Briefly describing the competitive landscape for VoIP, Mr. Mamakos observed that while VoIP is built around standards that allow devices to interact with each other, operators have built their networks using different back-office technologies. His customers' voice over IP adapter uses SIP to talk to his sibling system, the implementation of which represents an opportunity for Vonage to innovate and thus to compete in robustness, features, and other aspects of the service. One concern that the company has, in light of the rapid growth of demand for its services as reflected in both calls and minutes (see Figure 39), is the extent to which its system is scalable: "We don't want to experience a success disaster, with more customers than we can handle."

FCC Ruling Favors Innovative Telephony

Mr. Mamakos then turned to a November 2004 ruling by the FCC in favor of a petition filed by Vonage requesting that service of the type it provides be declared an interstate information service subject to regulation on a federal rather than a state-by-state basis. The burden of having to go to 50 or 51 jurisdictions with different regulations could significantly burden a new technology, he said, and would serve "only to delay introduction of new services and features." Remaining to be resolved, he acknowledged, were some issues of public policy, as many programs are funded directly or indirectly through taxes and fees associated with telephone services, and state and local governments see new entrants as causing erosion of such revenue. While expressing the wish to be "part of the solution, not part of the problem," he advocated solving the problem in a "uniform way" that would limit the burden of compliance to the greatest extent possible.

In conclusion, Mr. Mamakos declared voice over IP "here to stay" and posted the following claims:

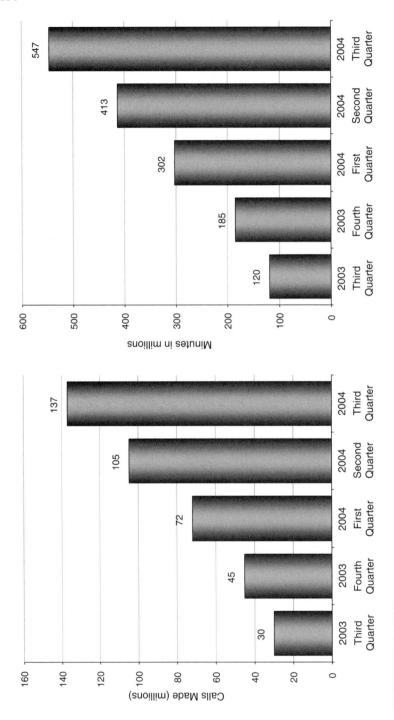

FIGURE 39 Scalability.

- that, as a "disruptive technology," it "has irrevocably changed the tele-communications landscape";
- that "pricing pressure" it has exercised "continues to challenge the voice-services models [that] incumbents have relied upon for centuries"; and
- that it is "challenging the regulatory framework, forcing new thinking with regard to subsidies, E911, and packet prioritization."

Offering an example of the Internet's ability to render geography "irrelevant to a large extent," he recounted that he had transferred his home phone service from Verizon to Vonage and taken his Area Code 301 phone number with him when moving not long before from the Washington, D.C., area to New Jersey. Just as any other Internet-based service, he said, VoIP is "inherently nomadic," meaning that where the user plugs into the network is more or less irrelevant, subject to performance concerns.

Presenting Andy Schuon, the president & CEO of the International Music Feed television network, as the next speaker, Dr. Murray remarked that the panel's remaining talks would focus not only on the interesting new digital content that can be provided by new technology, but also on difficulties in the industry that have come with the new technology's availability. Similarities were to be found, she said, in the music industry, the video industry, and the publishing industry, specifically with reference to digital books.

DIGITAL ENTERTAINMENT

Andrew Schuon
International Music Feed

As someone whose entire career had been spent on "the experience side of entertainment," Mr. Schuon said, he had been asked to talk about the content being delivered by the technologies that other speakers had discussed. He would therefore look at how music and other forms of entertainment were undergoing a fundamental shift in the digital age; at the effects of this shift on both the entertainment industry and the consumer; and at what needed to be done about the "widespread epidemic of piracy" that had accompanied it. He would touch upon satellite radio and video along the way.

Figures for legitimate downloading placed sales by iTunes at a couple of hundred million downloads and indicated that more and more people were using Napster and subscription services. But there were estimates according to which, Mr. Schuon said, legitimate downloading was running at only a few percent of all the downloads from the Internet. So while "a nice start" had been made, he said, "we still have a really long way to go."

Monetization a Key Problem for Content Providers

The public's desire to consume music had never been greater, observed Mr. Schuon, pointing to robust sales of portable music devices, some of which allow users to take an entire music collection with them anywhere they go. "People are falling in love with music again," he stated. "The only issue is: How do we monetize it?"

Technology developed for building legitimate services had in fact made it possible to protect intellectual property, to monetize it, and to track licenses while at the same time creating a good experience for the consumer. But the entertainment industry had been forced to wait for technology to catch up with its needs, and now it had to find a way to get the consumer to pay for its product while at the same time being more creative than the illegitimate download sites. Success would require what Mr. Schuon characterized as "a multi-pronged attack": Not only did the legitimate services have to offer as great a variety of songs as the competition but they also had to remember their obligation, being in show business, to entertain consumers. "We have to give them a dazzling product that isn't just high speed and fair priced but also offers them the flexibility to use the content in the way they really want to," he stated.

This, rather than price, was the issue. Indications were that those willing to pay for downloads found the range of 75 cents to a dollar per track to be in the ballpark, but the consumer wanted to take control. And rules had been associated with the use of content obtained through legitimate services, whereas "if you steal the content, you can do anything you want with it—put it on any portable device, put it on as many computers as you have, use that content as you see fit," he said. "That's what it's all about in the future of entertainment."

Technology Boosts Legitimate Providers' Offerings

With the advance of technology, however, this would become "much less an issue" for the legitimate providers. Mr. Schuon offered a recent iTunes release as a good example of a product "far superior" to anything obtainable through illegitimate sites: A "great experience and value proposition" that was well marketed and well priced, this was a virtual boxed set containing all 400 of U2's songs, including the band's new album and 30 songs that had never before been released. Priced at $149, it could be downloaded at high speed with one click, and users could put it onto a portable device such as an iPod, burn it onto a CD, and put it onto any number of computers. With the purchase of a U2 iPod, the set's price fell to $100.

"What we've been combating so far is trying to put digital rights management [DRM] onto all of these services, whether they be music or movies," Mr. Schuon said. Since the MP3 open standard is much easier to proliferate, the industry has had to wait for the technology to catch up. Although he conceded that DRM would never be foolproof, he said that it could "dramatically reduce the problem [while providing] those willing to pay a really great experience." Digital watermarking was also likely to offer a future alternative for tracking and providing protection. He cited the success that Verance Corporation had enjoyed in the course of 2004 in tracking all advertising on television and radio; noting that the company was moving over to the music space, he predicted its technology would be effective there as well. Microsoft's Janus project, also showing promise, would allow users with subscription services to take content and move it over to a portable device. Rather than pay $149 for the U2 package, some might prefer to "take the whole world of music with them for $10 [or] $12 a month," with their access to the content continuing as long as they subscribed.

Music: the Entertainment Industry's Guinea Pig

Mr. Schuon portrayed the music industry as a guinea pig that had been obliged to "take its hits" while business models, rights issues, technology, and other aspects of the new, digital entertainment market were getting ironed out. "It's paving the way for new experiences across the spectrum of devices: set-top boxes, phones, and all of video," he stated. Viruses, file spoofing, fake files, and "other things that lurk around in that third-world bazaar of illegal downloading" would continue to cause concern but would become less relevant as time went on. He praised as constructive the aggressive actions being taken by record companies, the Recording Industry Association of America (RIAA), and the Motion Picture Association of America (MPAA).

Technology, he stressed, was creating a new entertainment economy by giving consumers the control they wanted. For those with a TiVo, a favorite television program is on at whatever time they want. More and more, it is consumers who make prime-time television schedules, who make CDs, who make their own private radio stations by programming their iPods and taking their music with them. "It's very rapidly becoming absurd to think about going to a TV schedule to see when something's on or buying a CD that someone else chose to put a certain set of songs onto," Mr. Schuon said. Already, the technology available in mass-market consumer electronics stores would allow anyone to make his or her own multimedia network.

As a result, music and TV would in the future be marketed more like films. Content providers would no longer be able to rely as they once had on one show's leading into another or on a listener's waiting to hear a particular song on the radio. Distribution, as had been seen with the Internet, would no longer be in the hands of the major media companies; consumers would make their own content, and they would want to use copyrighted materials to really express themselves. "Legitimate peer-to-peer services are going to need to come to market," Mr. Schuon said. "Sharing and recommendations from friends will be essential and will drive the future of entertainment."

Content Licensed to, Originated by Consumers

Avenues would have to be found for licensing content to consumers because consumers would pass content around among themselves in any case. Mr. Schuon recounted that his niece, a day after he had read a bedtime story to her over a video chat link that spanned the continent, had phoned to ask him "when it would be on again." His comment: "There I am, creating content. It's basically my own Disney Channel for my niece when I'm out of town." While AOL, Yahoo!, Google, and a few others might emerge as the major aggregators of content in the future, the capacity to offer such choices as that exemplified by his on-demand bedtime-story channel was not far off. A new service that delivered television programming over the Internet via a set-top box, Akimbo, allowed the user to search for content on a given subject. "If you search for something on skateboarding, for example, you might get ESPN2 right alongside a show that anyone in the audience might have created," he explained, just as a Google search for a product might deliver offerings from a top-line merchandiser and a mom-and-pop shop right alongside one another and without the user's being able to tell the difference between the sources.

He offered another example: RipeTV, a video-on-demand channel supported by advertising that had been the subject of a recent magazine article. Founded by "a couple of guys who were bored, had made a bunch of money, and decided they wanted a TV channel," the venture was creating its own content and had made a development deal with Comcast Digital. While he was not convinced that the channel, thanks in part to wireless, would fulfill its owners' prediction of garnering 300 million viewers worldwide by 2006, Mr. Schuon was reluctant to rule out the possibility of success. "Really, content is king," he said, "and if you're out there creating these new forms of content on your own low-cost networks, there'll be more and more places to display that content."

'If You Don't Build It, They Can't Come'

Mr. Schuon's current company was active in launching cable channels, working with satellite radio, doing television production, and advising companies in the field of broadband entertainment. Calling himself "bullish on the idea of entertainment brands and content offerings," he said that he had often been asked to launch new television channels and had invariably accepted. "'Even if we don't have a lot of distribution right now, build that brand,'" he tells clients, because "'if you don't build it, they can't come—and there'll be lots of places for your content in the future.'" Remarking that it takes only a single show to create buzz, he said that "Trading Spaces" had made The Learning Channel a "household-name network" and that The Bravo Channel had had few viewers before airing "Queer Eye for the Straight Guy." Even animated cartoons on the Internet, such as the political satire offered at JibJab.com before the 2004 presidential election, had the power to attract millions upon millions of viewers. Accessed through a link that had been passed around "virally" through email, JibJab might in the future turn into a network seen via an Akimbo box or on a cell phone.

The key features of future search technology would be high speed, massive storage, and "fabulously elegant" user interfaces. Mr. Schuon envisioned fiber, cable, satellite, and wireless all flowing seamlessly into set-top boxes or similar devices equipped with ample storage incorporating search capability that would allow users to look at everything on their drive and "everything that [they] might possibly want to download or watch in the future." The box was likely to be leased, much of the content either rented or obtained under subscription. Current access providers—ISPs, as well as cable, satellite, and wireless providers—would, in his estimation, probably the best suited to becoming the "toll-takers" for users in the future. They had an established billing relationship with their customers and were set up so that they would be able to do the microbilling that would be needed whether subscription or a la carte became the dominant payment modality.

Owing to the proliferation of WiFi and WiMAX, satellite radio and radio services on broadband offering hundreds of channels would be ubiquitous in the future. Just becoming available was the XM MyFi portable player, a device the size of a deck of cards that, in offering all of XM's programming without a dish, was in essence putting "satellite in your hand." Its TiVo-like features enabled the recording of five hours of programming with rewind and fast-forward. "Good-bye WTOP," said Mr. Schuon, referring to a news-talk AM station popular in the D.C. area. "You won't need to wait for 'traffic and weather together' every ten minutes" because on satellite radio dedicated weather and traffic channels will operate 24 hours a day. He also predicted the disappearance of carpentry jobs "because there's going to be no need to build any bookcases."

Coming Soon: Changes Unparalleled in Profundity

According to Mr. Schuon, the ways in which content is made, marketed, distributed, and monetized—in fact, every element of the entertainment business—would change, and the change would be unparalleled in its profundity. Despite the many changes occurring in the automobile industry in its century of existence, its products still have four rubber tires and travel down a road. Even leaving aside the technologies used in distributing entertainment, he suggested, the number of platforms that content can be delivered on, and the variety of production, sales, and marketing techniques that exist, can sometimes seem overwhelming.

Looking back, Mr. Schuon said that his career path, one that had never strayed from the entertainment business, would not have been open to him if it hadn't been for a healthy industry able to monetize content and pay artists a proper royalty for their rights. At the same time, he said, he believed firmly "in the fan taking over the experience," adding: "The revolution will be televised, and it will be televised however the consumer wants to televise it to his or her friends, [and] on whatever platform they want" to use. In conclusion, he said that he planned to take part in ensuring that the field's "exciting evolution" continued.

SERVING CONSUMERS ON BROADBAND

Lisa A. Hook
AOL Broadband (retired)

Ms. Hook, having entered retirement only days before the symposium, put the attendees on notice at the outset that her style of presentation would reflect her new, relaxed frame of mind. She began by describing a longtime reluctance on America Online's part to acknowledge that broadband offerings would play a significant role in the consumer market for Internet services. "Historically the company has had a commanding market share in the dial-up space," she said, but it "also had a commanding ability to ignore the advent of broadband." It was only after some 15 million U.S. households had become broadband customers that AOL "decided maybe it wasn't just an early-adopter, propeller-head type of a product and [the company] should start paying a little bit of attention to it."

Its response was to import its business model for dial-up service into the broadband business. The dial-up model was brilliant and permitted AOL, according to Ms. Hook's description, to act as a "buying club" for internet connectivity: "We got a bunch of subscribers in the door, we went out and bought network connectivity—thanks to the 1996 Telecom Act, we could buy it more and more cheaply—and so our EBITDA [earnings before interest, taxes, depreciation, and amortization] margins were effectively driven by our ability to be the world's largest acquirer of network connectivity." In the dial-up space, this had been successful to the point of genius.

Replicating the Dial-Up Model in Broadband

Bent on replicating the model in the broadband space, AOL attempted to negotiate wholesale connectivity purchases with both cable operators and DSL providers. The former showed extreme reticence, resisting the company's entreaties that they open up their networks in a regulated fashion for reasons that were obvious. But AOL did conclude deals with the latter for the portion of the network that it needed, obtaining "great" line charges. It backed all traffic to its headquarters in Dulles, Virginia, put it through an "enormous amount" of processing, and sent it back out of its server architecture on the other side. AOL ended up providing "the slowest broadband service in the world," she said, adding: "The DSL guys were probably laughing all the way to their operations meetings."

Moreover, with respect to its own operations, AOL had unwittingly gone into a business that had nothing to do with its dial-up business. "On the dial-up side, we were able to acquire network connectivity and to handle the customer care into networks that had a high level of visibility down to the home," Ms. Hook explained. "In the broadband area, there's absolutely no visibility into the networks, so we were getting customer care calls and, frankly, not having answers the customers needed—never a good recipe for customer satisfaction." To complicate matters further, the company was warehousing more than 20 SKUs of DSL modems—something that those with experience in operations might recognize as "a very bad thing." To top it all off, AOL was managing its modem inventory next to that of another business it had: selling linens, seed pearls, and other such items. "They were all in these bins, and sometimes we'd send a DSL customer sheets and towels instead of a modem," she recalled. "Very difficult to get connectivity, even at 200 thread count."

Separating the Network and Service Layers

In sum, AOL was trying to force itself into a connectivity business in which it did not belong, but it continued on for some time—"losing EBITDA on an operating basis on every single subscriber [it] brought onto the network"—before taking stock of the situation. The company then decided to leave aside the network layer of the business, which it judged to be beyond its area of expertise, and to focus instead on the service layer: on developing innovative products, integrating them, and selling them to consumers under the AOL brand. "While we all now assume that the split between the network layer and the service layer has been out there for a number of years," Ms. Hook remarked, when this decision was made two years before, "it was quite revolutionary. Wall Street thought—as did some people inside our company—that we had lost our minds."

But the consumer, faced with a proliferation of Internet services, operating systems, and devices, still wants service that is both easy to use and integrated. This is true even of the early adopter, Ms. Hook asserted. For this reason, the AOL brand positioning of "simple and easy to use" was one that could be spun to the service layer without much difficulty. As a result, around 5 million users had signed up for AOL's broadband service layer at $15 and $25 per month in the previous two years, and 3 million more had opted for its premium services, which included voice, wireless, safety and security applications at $3 to $5 per month.

Point Service Explosion to Renew Demand for Aggregators

The market was thus clearly present at the service layer, concluded Ms. Hook, predicting future offerings of point services in advanced communications, as foreshadowed by Vonage, and entertainment, where the potentially "explosive" video over IP would be joining music. All such products would need to be integrated with each other, including those in the field of safety and security, which she rated as the preeminent market: "What we see people saying to us is, 'I've got a firewall, I've got anti-virus, I've got the spyware protection, but I'm dying here. Can you put it altogether so that I have one click into my system?'" The paid services of providers like AOL have and will become more relevant. In short, the proliferation of point services and of theme packages could be expected to lead back to the need for aggregators like AOL.

Companies participating in this service layer will have to get innovations to market very quickly, said Ms. Hook, pointing to dramatically shortening innovation cycles and to problems experienced by MSN in kicking off its Longhorn line of Internet products, as well as to similar problems at AOL. What this accelerated pace will require from large firms like AOL is "moving from our old mentality of building proprietary networks and systems to an open-platform type of architecture, and recognizing that our value add is in the brand, the distribution platform, and the customer care and billing on the back end," she said, adding: "People like us just cannot innovate so, like other companies our size, we are moving out and embracing third-party innovators."

Launching Applications as If They Were TV Programs

Over the previous year AOL had already taken advantage of shifts in the market to begin opening up its subsystems, inviting third-party developers to work with it, and then launching their innovative applications into the market in the way a television network would launch a program. "We put things up, we try them, we see whether consumers like them, we take them down if they don't, we put more marketing dollars behind them and integrate them into our service if they do," Ms. Hook explained. As Internet services moved off the PC and onto stereo systems, television sets, game boys, PlayStation 2s, and cellular services, the necessity of this would only grow.

To parry potential questions as to whether there remained a role for an aggregator such as AOL or Yahoo!, Ms. Hook said that while early adopters might be expected to share the symposium audience's level of sophistication regarding the Internet, members of the average user base would not. She recounted a customer-service call of a few weeks before, saying that she had made a practice of listening in on such calls to remind herself that people could experience problems with even the simplest of services. This call concerned a system that AOL offered to both broadband and dial-up customers permitting parents to set the level of access their children would have to the Internet. "We had one long-time customer call, and he was trying to figure out how to turn out the parental controls," she recounted. "He unfortunately was down in his laundry room, and he thought that the parental-control switch was near the boiler." Simplicity is needed, she declared, "so we don't have everybody who's trying to use these services wandering around the laundry room or worse."

THE VIEW FROM THE COPYRIGHT INDUSTRY

Steven J. Metalitz
Smith & Metalitz

Mr. Metalitz began by listing products and services that are dependent upon copyright protection: books, music and sound recordings, movies, audio-visual, TV, video games, computer games, and business software, among others. To illustrate the economic impact of the industries that produce them, he posted a chart showing the results of a study commissioned periodically by the International Intellectual Property Alliance, which he represents (see Figure 40). The most recent study, based on data for the year 2002, put the annual contribution of the copyright industries to U.S. GDP at $1.25 trillion dollars. Half of that came from the "core copyright industries," those he had just mentioned; the rest came from other, "copyright-dependent industries" including the segments of the retail, transportation, and distribution businesses devoted to copyrighted materials. Similar studies, of which more and more were being conducted, put results for other countries in basically the same range.

Pirate Product Inevitable with Broadband

Expressing his enthusiasm for the opportunities broadband affords to "everything that is protected by copyright"—opportunities to provide new types of products and services to new customers over new delivery media—Mr. Metalitz

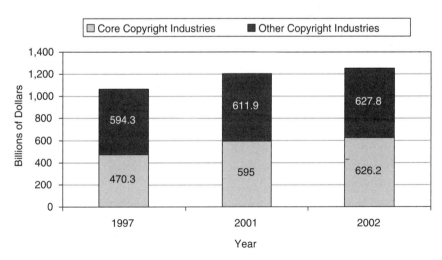

FIGURE 40 Copyright industries (ISIC) value-added contribution to GDP.
SOURCE: *<http://www.iipa.com/copyright_us_economy.html>*.

said he looked forward to the growth of broadband's presence in the United States. At the same time, he cautioned, "we know there's going to be a certain amount of pirate product coming through the pipe, and anybody who tells you that there's any realistic strategy to eliminate piracy on the network is fooling themselves or attempting to fool you." The hope, he said, was to achieve a relatively low level of piracy and a very high level of legitimate products; the concern, of course, was that the exact opposite might result. The broadband challenge was "to make sure that it's the first scenario, and not the second," that prevailed.

Referring to Dr. Nelson's observation that Napster, though demonstrating the power of a new paradigm, had nonetheless been illegal, Mr. Metalitz went further. "Not only is it illegal," he declared, "but it's also bad for this huge segment of the economy that we've been talking about. It's therefore bad for our overall economy, it's bad for jobs in the United States, and it's certainly bad for the public as a whole in terms of the continuing incentive to invest in the creation of new audio, video, software, and other products."

Korean, U.S. Broadband Markets Diametrically Opposed

Mr. Metalitz evoked trends from the music industry in South Korea—whose present, he suggested, may provide a glimpse of the United States' future—to underline his concern. Close to 80 percent of South Korean households have broadband access, a rate twice that of the United States, and the network is used differently in the Korean market than it is here. Four-fifths of Korean broadband customers reported consuming audio and video products, over half play games online, and some two-fifths engage in file sharing, while only 14 percent reported using their broadband connection for email. U.S. figures were close to opposite, with a far higher percentage of Americans using broadband for email, a far lower percentage for some of the other applications. In the music industry, whose role as guinea pig he ascribed to its needing less bandwidth than video, the Korean market for compact disks was off 40–60 percent from a few years before, to the point that it was smaller than the market for ringtones. "It's great that mobile services are growing," he reflected, "but that isn't really going to replace the much-larger hard goods market." The country had, at the same time, seen a huge increase in pirate services. When the Korean version of Napster 1, Soribada, was shut down, it had 8 million subscribers, or about one-sixth of the country's population. An unlicensed audio streaming service had 14 million. "We don't want to end up in this situation," he said.

But how might the United States avoid it? Mr. Metalitz organized the elements of the challenge under five rubrics: legitimate market, technology, legal tools, enforcement, and public education.

Legitimate market

Developing the legitimate market for copyrighted materials over broadband—for entertainment services, software, video games, research, reference works—was indispensable for success. Meeting this challenge would mean offering enhanced products, as had been done in hard goods with the transition from the CD to the DVD and on to the various types of enhanced formats whose presence in the market was increasing; offering more delivery channels; and making services easier to use. This new broadband market was analogous in certain ways to a large, new geographical market: He drew a parallel between it and the Chinese market, which the copyright industries had been trying to reach with physical goods and where they had encountered a significant piracy problem. While remarking that many steps had been taken to combat piracy in China, Mr. Metalitz suggested it was generally recognized "that you can't really change the paradigm and move to a society that's mostly consuming legitimate goods unless you have access to that market and can get your legitimate goods in." The widespread availability of infringing product represents a similar "market access barrier" for legitimate copyright industries in the broadband market. The key to surmounting that barrier is to make sure that the technology was married to the creative product in a way that delivered something customers would want and would find both easy to use and attractive.

Technology

Greater control over content would need to be provided to the end user, as Mr. Schuon and others had discussed, but delivery of the content would have to be sufficiently secure "to keep honest people honest." Also needed would be measures ensuring that the income-generating potential of material going into the pipe did not vanish forever. In addition, according to Mr. Metalitz, a more platform-neutral approach was called for; the problem of interoperability, salient in the music sphere, where legitimate services were proliferating, had not yet been solved. Finally, it was desirable that protections be developed that, when applied, were more or less invisible to end users—who could thus focus on enjoying the experience that they had paid for, or on getting access to the material that they had subscribed to, rather than on the protections.

Legal Issues

In the Digital Millennium Copyright Act, the United States had the basic framework needed to protect the technological measures used to control access to copyrighted materials in the network environment, and more and more countries were adopting similar measures. Some enforcement improvements were pending before Congress even then: The Intellectual Property Protection Act contained an amalgam of ways to improve enforcement activities, including those of the Department of Justice. But a problem had been presented by a recent decision of the Ninth Circuit Court of Appeals dealing with peer-to-peer services and, particularly, with Grokster. Under that precedent, according to Mr. Metalitz, a business could be built whose only viability was based on copyright infringement, "and yet that doesn't attract liability under the copyright law or, really, any other law at this point." This needed to be fixed, he said, because investment and innovation should be going into legal businesses rather than into encouraging illegal activity, and there were many ways in which it could be fixed. A petition was then pending before the U.S. Supreme Court asking that the case be reviewed, and legislation had been proposed in the Congress that would address the matter, although it was unlikely to be passed during the current session. However it was addressed, he said, the status quo was untenable because, under it, investment was permitted that encouraged illegal activity.

Enforcement

Sometimes forgotten "in all the brouhaha about the lawsuits that the RIAA [had] brought and that the MPAA [was] about to bring against end-user file sharers," Mr. Metalitz said, was that "most of the piracy problem we face is still due to organized criminal groups." Many of these groups are transnational, and that, he felt, is where many enforcement resources need to be focused. Still, what he called "dedicated amateurs" also played a role in making the system insecure, which explained the RIAA's and MPAA's actions.

Public Education

Enforcement action, however, was not being undertaken exclusively for its own sake; in fact, Mr. Metalitz asserted that it was best viewed as a means of public education. Survey research had shown that most people in the United States had not known a year or two before that uploading through a file-sharing service was illegal and an infringement of copyright if one did not have permission from the copyright owner. Those present at the symposium would have known that this activity was illegal, he stated, because they followed such issues closely, but the average consumer really had not been aware. That had since changed, in that most people now knew that it was illegal. The question of whether they were able to make their own conduct or that of their family members conform to the law, however, got into a cultural issue regarding attitudes toward intellectual property and creativity in the larger society. Nonetheless, public understanding, at least of this basic feature of copyright law, had moved to a much higher level.

To Meet the Challenge, Cooperation a Must

Possibly the most important concept for the copyright industry as it attempted to meet the broadband challenge, according to Mr. Metalitz, was cooperation. The copyright industry was unlikely to achieve or even to advance its objectives in any of the above areas, he said, in the absence of cooperation with providers of networking services on the one hand, and, on the other, better communication with policymakers and people such as those attending the symposium who were "seeking to understand what all these developments are leading to."

DISCUSSION

Philippe Webre of the Congressional Budget Office noted that Cisco's claim to have sold "a couple of million" IP telephones contrasted with Mr. Mamakos's estimate that, at 300,000, Vonage held half the VoIP market. He conjectured that many IP phones were being used in enterprises and asked Mr. Mamakos to explain how the enterprise market was different from the market he had discussed in his talk.

Mr. Mamakos suggested that the VoIP enterprise market might be considered "sort of a next-generation PBX." He himself had a Cisco VoIP phone on his desk at Vonage, and many other companies had abandoned the traditional Nortel/Avaya key system phone that is plugged into a central PBX using dedicated wires. Replacing it were voice over IP appliances bought from Cisco or other vendors; these are equipped with a "soft PBX" that handles VoIP and then is connected to the PSTN "behind the scenes by whatever means: either voice over IP again, or ISDN PRI, or the normal sort of interconnect." The statistics in his own presentation had referred to end-user subscribers on the PSTN as opposed to users within an enterprise or corporate setting. He said he did not know whether the enterprise market was growing faster than the end-user market.

VoIP Consumer, Enterprise Market Rising in Tandem

Dr. Nelson, expressing IBM's excitement about enterprise voice over IP, posited that growth in the two markets was similar, with curves going up very quickly. He stressed, however, that the enterprise sector comprised IP services beyond voice. It was the ability to put fax, email, and voice together in one system that IBM was selling to its enterprise customers, who were tired of managing separate phone and data networks. This was particularly useful to companies with large numbers of mobile employees; for example, it allowed IBM employees, 30 percent of whom did not have an office, to get their voice mail and faxes as email attachments. As a result, he remarked, "we don't have to go to three different services to get the information we need to do our job." So while Vonage was selling more versatile voice service, IBM was selling a totally different way of doing messaging.

Vonage, stated Mr. Mamakos, offered the same sort of technology and some overlapping products and services, but was targeting different market segments.

New Services: Identifying the Showstoppers

Evoking the term "showstoppers," used by the semiconductor industry for potential obstacles to the continuation of progress at the pace described by Moore's Law, Dr. Charles Wessner of the STEP Board asked Dr. Nelson and Mr. Mamakos whether they were aware of potential showstoppers of either a regulatory or technical nature in the markets they had discussed. He asked in addition whether solutions to any such barriers had been identified.

Dr. Nelson named privacy, intellectual property, and security as the three showstopper issues for The Grid, with a solution to the last being a prerequisite for dealing with the first two. As to privacy, a "major change in mindset" was needed before corporations would accept a third party's running the IT infrastructure on which their essential services depended; among other things, they had to be convinced that their data would be kept confidential even though the third party was running all the systems that that data passed through. "You not only need to make sure the data is safe from hackers," he said, "you also have to convince the corporate customer that you're not somehow tracking what kind of applications they're using and who they're talking to." This was similarly true, he observed, in the case of Vonage's customers, to whom it must be abundantly clear that the company was storing a great deal of personal data, voice mails included. The challenge for both grid and VoIP providers was to win the customer's trust.

On the issue of intellectual property, Dr. Nelson said, there was "a long way to go." No consistent standards for DRM existed, and it was clear that many different solutions would be needed, as The Grid would be a very powerful tool for the pirates referred to by Mr. Metalitz.

Imposing Old Regulation on a New Medium

Dr. Nelson then turned to security, which he considered not a regulatory issue but a challenge for suppliers of services, as hundred-billion-dollar industries were being built on an infrastructure that "isn't quite ready for that." In the regulatory domain, his biggest fear was of efforts to impose old regulation on the new medium—a tendency that was, in fact, active all around the world. Serving as vice president for policy for the Internet Society and working with developing countries, he had seen how easy it was to say, "Internet telephony looks like telephony so we'd better to regulate it that way, and streaming audio looks like radio so we'd better impose regulation that way." Such thinking, he asserted, "could stop everything very quickly."

Mr. Mamakos placed the major challenge for Vonage in the domain of regulation and public policy. There was "no new physics we have to invent to be able to grow the business," he said. "It's a matter of doing a good job of execution, and technology evolution hopefully allows us to do a better job more economically." In contrast, great uncertainty remained on the policy and regulatory side. Vonage was engaged in a project to implement CALEA capability, as the company wanted to be "part of the solution, not part of the public policy problem." While his own belief was that the company needed to pursue the project even though great expense was involved, he acknowledged that this was "not entirely clear."

If the grid is inherently global, Dr. Murray then asked, "what do we do in the United States?"

Locating Transactions in an Inherently Global Market

Mr. Mamakos remarked that Vonage's users could, and did, take their telephone adapter and plug it in all over the world, using their telephone service as if they were at home. "It's a feature," he said, "not a bug."

Dr. Nelson observed that while many old laws assume that a company is located in one place and that a transaction occurs in one place, The Grid might have data coming from Brazil and computing power from Canada and Germany while the user was somewhere in Belgium. Thus the potential existed "to do a transaction in five places at once," which would also confound regulations taxing the value of the transaction based on where it took place. That the "laws are not virtualized but The Grid is" could lead to what he termed "a lot of just total collision."

Mr. Metalitz said that all three issues raised by Dr. Nelson were problematic on the international level. That legal standards were harmonized to a greater degree than they had been a decade or two before in the intellectual property areas, and particularly in the copyright area, was a positive development. But there was far less harmonization in the privacy area, where there was not any one international agreement that established a standard of privacy protection.

Ms. Hook, while concurring on the previous points regarding IP and privacy, differed on taxes and subsidization. "It is really quite easy," she contended, "for service-layer providers to provide the services from an international point of presence and avoid having to get at all entangled in any kind of local or state taxation or universal service fund." This was particularly important on the telecom side, where upward of 23 percent of the revenue line was subject to a variety of state and local taxes, making domestic providers "automatically non-competitive vis-à-vis providers coming in from international sources."

Technology's Challenge to the Constancy of Time

Mr. Hellman, in reflecting on how the evolution of computer and communication technologies might affect real estate development, had realized that people live simultaneously in two different domains, time and space. Asking someone the distance between home and work, for instance, more often elicits an answer expressed in time than in space. Examining the reason people respond this way leads to a profound insight: There have always been 24 hours in a day, and in the future there will presumably continue to be 24 hours in a day, and because all basic activities have to fit into 24 hours, people's behavior is relatively stable in the time domain. But as work becomes more virtual, with technology allowing it to move away from a paper-based manual-labor paradigm toward an electronic-based network paradigm, the whole world becomes one virtual place in which Australia, say, is less than one-tenth of one second from the United States. The challenge for the evolution of the human species, in light of this, may be whether it can deal with the world as one integrated system.

Dr. Nelson remarked that one of the drivers behind The Grid was the desire, since there are only 24 hours in a day, not to waste half an hour of that time backing up a hard drive, reconfiguring a disk, or downloading new software patches.

Mr. Schuon recounted having experienced a "kind of content Moore's Law" over the previous two years thanks to TiVos and other products incorporating hard drives. He now flips through TV shows as through pages of a magazine, forwarding through parts of shows that do not interest him. "I can watch 'Dateline NBC' and '60 Minutes' in under and hour," he said, "and I'll bookmark the business section of the *New York Times*, reading that every morning but never getting to the rest of the paper." By managing content in this manner, he has been able to do and see much more in the same 24-hour period.

The IP Dilemma: Private Licenses vs. Public Good

Mr. Hellman observed that information is an unusual form of property in that even if it is stolen, the victim of the theft retains it. He saw in the current increase in the flow of intellectual property a parallel to the change brought about by the invention of the printing press, which had drastically reduced the cost of reproducing and distributing information, thereby elevating the quality of life on earth as a whole. The intellectual property issue becomes very interesting, in his view, if the only way to protect what we think of as intellectual property rights is to slow down an evolution whose benefit to the world is such that economics may be irrelevant.

Mr. Metalitz noted that intellectual property law had faced many challenges in the past and had adapted to significant changes. A century ago, for example, there was no such thing as recorded music, and copyright law has been able to adapt to all the changes that intervened. Still, he acknowledged, there was no question that, as Mr. Hellman had outlined, it was again facing a profound challenge.

Mr. Schuon, recalling Dr. Raduchel's description of how people gathered to listen to the virgin play of an LP, observed that the current consumer was largely happy with an MP3 file on an iPod, whose quality was well below that of a compact disk. Consumers, he concluded, would trade quality for the ability to manage content and use it in a more exciting way.

Participants' Roundtable

Moderator: Dale W. Jorgenson, Harvard University

H. Brian Thompson, iTown Communications
David S. Isenberg, Isen.com
Lisa A. Hook, AOL Broadband (retired)
Jeffrey M. Jaffe, Lucent Technologies
Andrew Schuon, International Music Feed
William J. Raduchel, Ruckus Network

Dr. Jorgenson asked each of the roundtable's participants to state briefly what he or she would be taking away from the day's meeting.

As someone who had repeatedly had the experience during his career in the telecommunications industry of looking back and seeing the preceding few years as the most exciting yet, Mr. Thompson said he had the feeling as the symposium concluded that the coming five years were going to be extremely exciting. Significant discontinuities would be taking place "in everything that everybody thought was the status quo," he predicted, quoting MCI's founder, Bill McGowan, on the danger of "wanting the status quo long after the quo had lost its status."

That those in industry desire both to lead commercially and to cause their technologies to evolve into something useful to society dictates that they look at what they are trying to accomplish from more than one perspective. Putting in a word of praise for open access, he said that "we've got to create the opportunity for people not to be shut off even from the small things that we find not just enjoyable but beneficial to us intellectually as well as socially."

Mr. Thompson found frustrating that a huge part of the world still had never used a telephone, let alone taken downloads on an iPod. He stressed the importance of opening up access to the network and "creating opportunity for the people, broadly based, to make their own selections about what they want to do." Endorsing Ms. Hook's plea for simplicity, he said that services needed to be not only accessible but also packaged in a way that would allow users throughout the nation and the world to take advantage of them.

The symposium had also, he said, left him with the pleasure of knowing that his was not a lone voice on such issues.

Mr. Isenberg asserted that "if the status quo has lost its status, people in general at this conference don't seem to realize it." Technology that already existed made it unnecessary to be limited by bandwidth, but those in attendance were "just assuming that we're on the right track." If, as others had claimed, the demand for bandwidth was insatiable and people would use bandwidth once they were given it, then the problem, as he expressed it, was that "nobody's giving me bandwidth." What he wanted, he said, was "not just bandwidth [but] stupid, open bandwidth: fast pipe, always on, get out of my way."

Ms. Hook recounted that in the summer of 1949 there was an article in *Electrical Engineering* magazine predicting that within the next three years all homes would be networked, with a server in the telephone closet. In the late 1970s and early 1980s Warner Communications founder Steve Ross had an interactive network called "Cube" in Columbus, Ohio, intended to deliver voice, video, and several additional applications, as well as to upload and download the user's medical information, in a networked household. It was, she said, "a complete fiasco." Time Warner had a similar full-service network about ten years later that provided voice, video, data, uploading, and downloading on a trial basis to several thousand homes in Orlando, Florida. A survey conducted upon the conclusion of the experiment found that the only application users would have been willing to pay for was a high-pitched tone that killed roaches. "So for $100,000 a home," she said, "we had created the networked version of Terminix."

Juxtaposing the picture emerging from this 50-year span of industry history with the current argument that if people are given unlimited bandwidth, they will be able to use it, Ms. Hook offered several observations:

- that, in line with what Mr. Thompson had said, a point of inflection had been reached with respect to the ability to put unlimited bandwidth into people's hands;
- that, particularly over the previous year, people had begun to experiment in the manner of Mr. Schuon with uploading and downloading, and to change their behaviors in ways never seen before; but

• that the industry was struggling at the service layer to find business models and revenue streams on these applications that would justify the investment needed to make unlimited bandwidth available.

"So I think we're getting there," she said, "but I'd still say we can't build out this entire network based on Terminix's business model, because they're doing just fine with the old juice."

Dr. Jaffe's strategy for summarizing the symposium, which he said had presented so many issues and perspectives that his head was spinning, was to take all of the day's ideas and put them into five different "buckets":

• **Bucket No. 1, technology issues.** Technology issues concerning investment, infrastructure, security, and reliability remained to be resolved before service could be offered that was universal, high speed, and full bandwidth with a large number of applications.

• **Bucket No. 2, regulatory policy.** All had concluded that it is not acceptable for the United States to be thirteenth in the world in broadband and that, "somehow or other, our regulatory policies are failing us"; therefore, regulatory policy needed to be addressed.

• **Bucket No. 3, economic policy.** There had been a great deal of very interesting discussion about who the winners and losers would be in the next generation of networks. "What's important to me," said Dr. Jaffe, "is that the consumer wins, that the country wins, and that we get this technology out there."

• **Bucket No. 4, technology absorption.** "Technology absorption by the masses is a non trivial part of the technology and industrial ecosystem," said Dr. Jaffe, praising Ms. Hook's points regarding the important role played by the service infrastructure.

• **Bucket No. 5, R&D policy.** While much of the symposium had been devoted to "how we deploy that which we already know how to do," Dr. Jaffe was anxious about "that which we don't know how to do" and about making sure that the United States got there first. The country had been very innovative over the past two decades, and the day's discussion had indicated that it would continue to be innovative over the coming two decades, but, he said, "we don't want to stop there." For this reason, R&D policy was, for him, a "hot-button issue."

Mr. Schuon evoked the memory of the "perfect storm" for entertainment content providers that had arisen from the confluence of a number of elements: the availability of ample space on hard drives and of a certain amount of broadband with the advent of MP3s and then of Napster. This ushered in an era of widespread piracy and file-sharing, and it made music the killer app driving much of the growth of the Internet's entertainment content sites in the late 1990s. At that time, every major media company thought it had to have music sites, as music sites were going to drive its business.

The music business, Mr. Schuon reiterated, is the guinea pig for other content businesses, and, with larger amounts of broadband becoming available, the movie business was now right behind the music business. Remarking that not all financial models had yet been figured out, he expressed the hope that the content sector of the entertainment industry could "catch up to the technology a little bit and get a little bit ahead of the game." Because providers had neglected to think about how to monetize content on the Internet, they were now having a hard time convincing consumers, who had become used to the idea that everything coming to them via the Internet was free, that they should pay. "If we don't educate people early about the value of the content," he said, "then when all [the anticipated new offerings become] available, everybody is going to be trading it again, and we're going to have no business at all in media."

Mr. Schuon called this a frightening prospect but said that, although it might seem to be just around the corner, it would be longer in coming than many—and, in particular, enthusiastic potential users—might think. Still, he warned, if the industry tarried in creating an economic model for "the distribution of content on a grand scale," then it would find itself "in very big trouble."

To conclude the panelists' accounts of what they would take away from the symposium, Dr. Jorgenson turned to Dr. Raduchel, who had set the context for the day's discussions by providing what Dr. Jorgenson called a "very eloquent picture of the convergence of everything on the great network."

Dr. Raduchel began by noting that Mr. LaJoie had been the only one among the network providers to raise the question of content protection, as well as by seconding Mr. Schuon's observation that content protection would be essential for the future of the entertainment business. Auguring a significant challenge was the fact that content protection had not been built into the network, nor had the industry yet come up with an alternative. His response to the question of "who was going to win, the broadcast industry feeding large hard drives on TiVo-like devices or the IP network-based people," was that "one side has all the licenses and the other has none." Seeing in this the existence of a "major edge," he said the question needed to be addressed.

But the foremost challenge, according to Dr. Raduchel, was coming up with a viable business model. He had recently returned from a visit to South Korea during which the president of Korea Telecom had told him: "'Bit consumption is going up all the time. Prices are flat to down.'" As a result, the company, although "way ahead" of its U.S. counterparts, was not making any money—or, as its president had put it, "'We're dying.'" And Messrs. Wegleitner, LaJoie, and Thompson had all just said the same thing: In the absence of a stable business model, there was no knowing when it would be possible to make a return on these activities. Mr. Thompson in particular was making "a huge bet" in positing payback over 10–12 years, which Dr. Raduchel called "a long time in today's world."

Convergence was coming, and every time it seemed that the flow of money into it might stop, it began again from yet another source. How would telecom companies deal with making cell phones work perfectly everywhere? with VoIP causing rates to plummet? with Skype, a company made up of eight computer programmers in Estonia, offering telephone calls worldwide for next to nothing? The next decade, Dr. Raduchel warned, would be marked by "lots of dislocation." Alluding to the remarks by Mr. Tenhula of the FCC, he placed the industry "somewhere between consumer confusion and political anxiety" and opined that the latter was just beginning to build.

Among his own worries, he said, was where the R&D would come from. Korea had no expectation of making money on consumer broadband but had built out its network because 38 percent of its exports were IP based and it wanted to make sure that Samsung and LG were the companies that supplied the world. It was clear that the network, as the Koreans did not care whether it ended up having value in itself, was an instrument of industrial policy, and that their interest was in driving investment, driving standards, and giving their home companies an advantage. It was because Dr. Raduchel felt that U.S. competitiveness hinged on broadband, even if he was unable to sketch the connection directly, that this was among his worries.

Although unsure what the business model for 100-Mbps-into-the-home might be, he was certain that it would depend on supplying content. Recalling that it was not market forces but, in the face of opposition from the cable industry, an act of Congress that got HBO onto satellite, Dr. Raduchel said he thought "it might take a similar act of Congress here around content." The European Union had been bolder, having overridden existing exclusivities in ordering that soccer rights be made available to broadband, both wireless and wired. "If you don't have the content, you're not going to have the business models to do this," he observed, reminding those in attendance that the reason consumers pay is to be entertained.

Summing up, he said he could not predict who would mediate the coming dislocation. Whether the FCC had the legitimacy to do this was unclear, as the FCC's moves to protect content had constituted "at best a mixed bag." There was, in fact, no one institution that was constituted such that it could settle questions of intellectual property rights. His initial question—how regulation would be achieved once convergence had turned all the various telecommunications industries into features—had thus remained unanswered.

DISCUSSION

Mr. Gardinier said that his major takeaway from the symposium was the Sturgeon Rule—"90 percent of everything is crap"—and inferred that it was exemplified by the practice of bundling. "People are being sold CDs when they want a song, they're being sold a DVD with 10 different applications that, because they just want to watch a movie, they never are going to use," he noted, asserting that this is "one of the big problems" in that it drives up prices. That said, he asked whether any of the panelists had taken consumer wants and needs into account in forming business models.

Mr. Thompson, emphasizing that he was speaking as a service provider and not a content provider, said that his own business model was based on the fact that "the consumer today is paying $137 a month to get something that's not worth $137." An assumption of his targeting a ten-year payback was that he could take out costs that service providers had been passing down to consumers. Consumers "go catatonic" upon reading their phone bill because none of the numerous charges it contains, the largest of which is the federal access charge, makes any sense to them. And it is these charges to which they are objecting.

Bundling was a second source of objection. "People want it simple," he said, "but they want it simple the way that they can consume it and the way they want to consume it." While having open access is difficult, it leads to the consumer's having more say. "It will get people who want to bundle in different ways for different audiences together to create something that's of great value to the individual," he claimed.

Addressing the question from the content side, Mr. Schuon said that since the American record industry had lost the ability to make money selling singles, which continued to be a decent business in Europe, it had been plagued by the bundling problem. Over the years, the U.S. industry had not done the best job of marketing the value of music to the consumer. Certain offerings could be considered complete works that actually constituted an album, he said, naming as examples certain classical works, movie soundtracks, and the music of such artists as U2 and Bruce Springsteen. But in most cases popular records were built around two or three songs, with the rest put in as filler. The fact that it had not figured out how to make money on an a la carte basis had caught up with the record industry now that consumers wishing to purchase single songs had the option of taking their business somewhere else, and it was something that the industry had to deal with.

DVDs, in contrast, were reasonably priced, and the extras were thrown in as a bonus, which consumers liked and which added value without raising cost. To illustrate how the pricing of DVDs had hurt the record business, Mr. Schuon recounted going to Best Buy and seeing the *Spiderman* DVD displayed right next to the *Spiderman* soundtrack and priced $2 below it. Such a thing is possible because there are so many marketing windows in the movie business: A film is sold at the box office, in stores, on pay-per-view, to airplane passengers, on HBO, and to broadcasters. In contrast, records have only the one play, they sell fewer units, and the cost is greater. For this reason, albums cost more than DVDs; consumers, however, have no idea why, whereas they know how much movies cost to make because "the film industry always tells you, 'this movie cost $100 million and this star got this much, and there's real value in that.'" While admitting that the record industry had not taken the time to educate the consumer on this point, he speculated that it might not be possible to do so in any case.

Comparing prices through the different distribution channels, Ms. Hook observed that it seemed fairly rational to pay $17 for the ownership of a film as opposed to $10 for a ticket to a movie theater. The problem for the music industry was that the alternative to paying $17.99 for a CD was not paying at all but stealing the content off the Internet. The issue, as she put it, was that the music industry had fallen behind in figuring out interim price points between zero and $18. In response, Mr. Schuon called charging a dollar per track "a nice start" but noted that the industry still can't make a profit at that price level.

Dr. Isenberg threw out two questions: whether Apple could be considered part of the music industry, and whether it hadn't figured out how to sell singles successfully.

Ms. Hook answered that singles were a loss leader for Apple the same way music was a loss leader for Wal-Mart. "This whole industry has become a promotional industry," she declared. But Mr. Schuon pointed out that the world's largest music company, the Universal Music Group, was not in the hardware business at all and therefore drew no benefit if its works promoted the sale of iPods or stereos—yet it still had, somehow, to turn a profit. As for the claim that iTunes was profitable, he speculated that if it had sold 150 million downloads at a dollar each, the business would emerge as "at best a breakeven" once Apple's marketing costs were figured in.

"Couldn't it be that the music industry is changing and that the formerly 'greatest music company in the world' is losing its status in the current quo?" Dr. Isenberg asked. Mr. Schuon replied that consumers had never had a more voracious appetite for music, and that there were more hits than ever, but that monetizing that enjoyment of the music had become extremely challenging.

In the opinion of Dr. Raduchel, iTunes in fact constituted "a disaster for the music industry" because, on a $330 sale of an iPod on iTunes, the music industry got $20 and Apple got $310. This business model was not sustainable for the industry in the long run because it would not provide sufficient revenue either to produce the music or to create the demand that drives sales. Moreover, as Steve Ballmer had said, the basic DRM for music on the iPod was "stolen." While the Microsoft CEO issued a retraction owing to the annoyance his statement had caused, according to Dr. Raduchel only 30 songs on an average iPod had actually been sold.

To address the question of bundling, he turned to the price structure in effect at McDonald's and the consumer behavior it engenders. A customer ordering a double cheeseburger, a coke, and French fries may base his or her decision on the price of the cheeseburger, but there is in fact no gross margin on that item, while there is a 90 percent gross margin on the coke and the fries. The company, therefore, depends on the sale of bundles for its profit. "The essence of every consumer business I've ever studied," Dr. Raduchel remarked, "is that you make your money by denying consumers what they most want, thus making them buy it the way you want them to." As this model was used in consumer businesses across the board, bundling would continue.

Observing that the symposium had spent far less time on the issue of security than did most meetings of its kind, Dr. Nelson posed a question on that topic in the form of a scenario. Some who knew the Internet well remained worried about the prospect of a systemic problem: perhaps a way in which the Domain Name System (DNS) could be brought down or a virus that, spreading throughout the 'Net, could disable a million hard drives in a day. How would the investment community respond to such an occurrence? How would regulators, both in the United States and elsewhere, respond? And how should they respond if the Internet has a serious security problem that could be exploited?

Dr. Jaffe expressed the belief that a single instance would occasion a great number of meetings and perhaps even a congressional investigation but "very minimal" response beyond that. The dearth of practical measures taken after the events of 9/11 had shown the country's first-responder systems to be antiquated could be regarded as precedent. He put forward as an explanation that in order to believe that there is economic value in defending against a security threat, "you have to believe it's going to happen again or build into your model that it's going to happen again." Because the United States is "a very optimistic society," he said, "it seems to take us a very long time to take things seriously." Agencies of the government that are nominally responsible for the country's infrastructure should be more active in this area than they had been, because nothing would happen otherwise.

Dr. Raduchel attributed recent resignations at the Department of Homeland Security to the lack of attention given this issue and said that Richard Clarke was very passionate about it but that, once he had left, there would be no one to carry the torch. As to whether a security disaster was likely to happen, Dr. Raduchel expressed uncertainty. He personally had been using the PC firewall product that had been voted the most secure, but attackers specifically targeting it had found a hole in it a few months before, with the result that his hard drive had been completely destroyed. Still, he registered his disagreement with Dr. Jaffe, saying that a major D-DOS attack would incite "panic in some quarters over what it would do and where it would go."

In the face of Dr. Jaffe's rejoinder that significant security breaches had already taken place, Dr. Raduchel objected there had been nothing "to the extent to which you couldn't make airline reservations for a week." But Dr. Jaffe held his ground, saying that incidents "pretty close to that scale" had occurred, including the shutting down of Amazon.

Ms. Hook drew a parallel to the major power outage of two summers before that had crippled the U.S. Northeast, among other things taking down all financial services and causing them to be rerouted to North Carolina. Although it was a "huge issue" at the time, she recalled, "nothing happened in Congress [and] nothing has happened since then [except for] a couple of days' worth of news stories." The incident would, in her judgment, have had to be "much, much more dramatic" for anyone to take real action.

Dr. Isenberg weighed in on the second half of Dr. Nelson's question, that addressing what should happen. First, arguing for alternatives to the DNS system, which he called "a single point of failure," he recommended a research effort that would investigate options. Second, he called for the institution both of "multiple, completely physically dissimilar ways of accessing the Internet" and of completely dissimilar Internet backbones transiting different parts of the country. Dr. Nelson mentioned that the Defense Department was in the process of implementing such a program, but Dr. Isenberg dismissed it because use would be limited to DOD itself. While he was unsure about advocating the resurrection of microwave transmission, Dr. Isenberg said that diverse fiber routes should "certainly" be made available. Third, it was important to have multiple and different operating systems: Having one operating system that accounted 85 percent or more of terminals was unacceptable from the point of view of network reliability.

Robert Hershey asked whether increasing bandwidth and computer power, the consequence of which seemed likely to be an improvement in capability by orders of magnitude, would simply make currently available services cheaper or lead to innovation.

Dr. Isenberg said this would lead to a world unimaginable even to those with active imaginations. "If we all had a gigabit to our home, which is probably affordable and within the scope of today's technology, life would be really different and, I have to believe, immeasurably better," he said.

Dr. Jaffe, recalling jibes from earlier in the day about the symposium's being a physical rather than a virtual meeting, confided that he "didn't relish waking up at 4 in the morning to go to the airport, pass through security, and board a plane in order to get here. Agreeing that the changes cited by Mr. Hershey would make a big difference, he predicted that "once there's enough bandwidth, there'll be this visceral feeling of being there even when you're not there."

Ms. Hook said that most bandwidth providers with whom AOL talks want to know, before they invest in the bandwidth, what applications it would be used to provide and how much people are willing to pay for them. She therefore called Mr. Hershey's question—which she rephrased as "What are they going to do with it?"—the question of the day. She asked Mr. Thompson to what degree he was willing to speculate: How extensively would he build out his network before his market could be sized and those willing to write monthly checks for his service had been identified?

Mr. Thompson emphasized that his project was focused on present rather than future capabilities. Having looked at what was happening in Japan with its $28 gigabits and at what was going on in Korea, he had become convinced that the crucial issue was not that of the technology's availability—rather, it was the cost of doing something with it that was beginning to drive new applications.

Had the United States moved toward more of an open network three or four years before, as in his judgment it should have, it would already have IP TV. That was the killer application that would "change the world as we know it," he said, adding that AT&T had thought the same in 1956, when it introduced the first television picture phone. Not only was Mr. Thompson himself convinced that IP TV would be coming along in the following two or three years, so were Southwest Bell, which was taking a big bet on it, and others as well. While admitting that how they would deliver was "another story," he maintained that lower costs were generating more opportunity to develop applications that could be useful. At present, the country was paying for capabilities that it was not taking advantage of.

Dr. Isenberg, however, "respectfully" disagreed that TV over IP had not yet arrived. He said that he often "went" to FCC meetings that put co-diffusing TV over IP, although the picture was, unfortunately in his view, both small and low-resolution. He objected to the notion, thrown out by Ms. Hook, that he was the only one interested in watching this, saying that specialized niches existed. He allowed as how the clips from *The Daily Show* that were being passed around were more popular than the FCC meetings, although he remarked: "If I had the bandwidth, I'd like to see that glint in [FCC Chairman Michael] Powell's eye when he says what he says."

Dr. Raduchel, looking at R&D trends, said that the one field in which the country was spending billions of dollars pretty much out of the public view was sensors. A large portion of the research was classified, as it was meant for application in such homeland security concerns as border protection and monitoring, but its results would inevitably flow into public use. Sensors could, in fact, emerge as the source of the next round of productivity improvements, and they had the potential to cause other significant changes as well. "If you put sensors every-where, you change the way that every logistics system works," he said, adding: "In the end, every business is a logistics system."

Pointing out that all sensors are silicon, Dr. Raduchel remarked that the de-clining cost of semiconductors was still "driving the world." And a great deal of bandwidth would be consumed by the constant communication among sensors, which don't do any good if they can't communicate. It would be above all the dumb sensors, those lacking the built-in intelligence to make decisions regarding the meaning of the information they were gathering, that would communicate in large volume. Asserting that the impact of sensors on the economy would be large—although its scope had not yet become clear, something also true of nanotechnology—he predicted that, within three to five years, they would provide the focus for many meetings similar to the day's symposium.

Concluding Remarks

Dale W. Jorgenson
Harvard University

Dr. Jorgenson said that while he did not wish to minimize the many interesting disagreements that had marked the day's proceedings, he felt justified in concluding that, in the course of them, lessons learned from previous symposia in the STEP Board's series on the New Economy had been corroborated.

The series, he recalled, had started with semiconductors and gone through computing and software before arriving at telecommunications. What is different about the last, as those attending had heard over and over again, is legacy. This legacy easily goes back to Alexander Graham Bell although, for economic historians like Dr. Jorgenson's neighbor, Al Chandler, the person most important to the industry's origin was Benjamin Franklin. Why Franklin? He was the first postmaster of the United States. But if communications is an industry with a huge legacy, it has certain features that are very similar to those of other industries that the conference series had examined, something Dr. Jorgenson would talk about in these closing remarks.

The first issue he would consider was economic impact. The telecommunications industry accounts, by various measures, for about 1 percent of the economy. But, Dr. Jorgenson cautioned, "'1 percent of the economy'" is the way industries are characterized not by economists but by the business press—so that, in *The Wall Street Journal*, an industry is accorded importance depending on its percentage of GDP in dollars. Economists, in contrast, look at an industry in terms of its impact on economic growth. And if the telecommunications industry were appraised from the latter point of view, he said, a rough estimate would be that this 1 percent of the GDP was responsible for about 10 percent of economic growth. Ten was a common multiplier; the New Economy, while only about 3 percent of the overall economy, was responsible for about 30 percent of economic growth, and the figure was rising. The corresponding figures for other countries trailed those for the United States but were on basically the same track. The economist's contribution was to get that multiplier of 10 front and center. "I realize we didn't totally succeed," he admitted, "but we tried."

The second issue concerned a standard industry model that emerges in the New Economy, a subject treated by Andy Grove in *Only the Paranoid Survive*. This "beautiful" book, which Grove wrote when he was not only CEO of Intel but a part-time business school professor as well, described a huge shift from a vertical model to a horizontal model in both the computer and semiconductor industries. In the vertical model—incarnated by IBM, which for a long time *was* the computer industry—there was a huge laboratory that created large portions of the intellectual property involved in the development of computing technology. But while such labs might still exist, the model itself was "history," as Grove's book had pointed out more than a decade before. What the day's discussion had taught, Dr. Jorgenson said, was that the identical shift had taken place in the communications industry. AT&T had been the IBM of that model, and R&D under it had been conducted at Bell Labs. In the new model, which had yet to be precisely defined, most of the interesting innovations were disruptive: Vonage was an example of a disruptive innovator, as was Skype, through which a mere eight programmers were disrupting the entire communications industry.

With the third issue, the business model, the most important question to be asked about any New Economy venture came to the fore: "How in heaven can we ever make any money out of this?" Agreeing with Dr. Raduchel that the way to make money is to figure out what the consumer wants and withhold it, Dr. Jorgenson commended the former's analysis of the cheeseburger-drink-French fries bundle's cost structure as an explication of the business model. Pointing to the existence of many books on the subject, of which his own favorite was Varian & Shapiro, he said the matter was very well understood and it applied, to a greater or lesser degree, to all of the businesses carrying the New Economy designation.

Despite the business model's familiarity, Dr. Jorgenson cautioned, it is hard to make money operating under it because consumers are both clever and unpredictable. It was "too bad," he said, that the consumer ends up carrying away most of the welfare, which then cannot be delivered to shareholders. But in another respect the fact that "consumers emerge over and over again as the big winners, even when confronted by tremendous intellectual talent, [was] the great thing about the New Economy."

This brought Dr. Jorgenson to the policy issues, which he qualified as extremely difficult. He characterized the economist's typical stance as: "'Private property has to be the answer, so we just create property rights and then leave the scene and let them fight it out.'" But in the telecommunications sector things are not quite that simple, because of the presence of other, related issues—common property rights, for example, or infrastructure that has to be maintained and compensated. So, while "it's not just a case of private property rules the world," a way had to be found of maintaining common facilities within a market-based approach.

In addition, there were more subtle property issues: How do you protect property? How do you keep it from being stolen? It was difficult to prevent consumers' not merely benefiting, but benefiting illegally and carrying off other people's property, which was obviously undesirable. At the same time, it was important to provide privacy in a convincing way, one that "assures people that they are in fact going to be able to enjoy their property in private if that's their wish." Property-rights questions spill over into other areas as well: hardware, infrastructure, software, and content as opposed to software. One very important area was the right to the R&D product, "the intellectual property that is created by people who are trying to create this new future that we've heard about here."

Dr. Jorgenson called the day's proceedings "a total success for the New Economy model" and credited the collective efforts of the presenters, panelists, organizers, and members of the audience who had participated in the discussions. He thanked all for attending and said he hoped they would be able to follow these issues as they unfolded.

In closing this meeting and the series on the New Economy, Dr. Jorgenson offered special thanks to Dr. Wessner, Dr. Shivakumar, Mr. Clabaugh, and Mr. Dierksheide of the STEP Board staff, and above all to Dr. Raduchel, whom he called the intellectual leader of the STEP Board's venture into the New Economy.

III

APPENDIXES

Appendix A

Biographies of Speakers*

MARK E. DOMS

Mark E. Doms is senior economist at the Federal Reserve Bank of San Francisco. His research interests include diffusion of IT technology and effects on firm performance and shifts in IT investment. He is currently beginning a rather large project examining the relationship between technology use and firm performance. This study will use a very large data set on technology use at the establishment level between 1980 and 2002. The study will first examine the diffusion of IT technologies over the past several decades, then examine the relationship between the adoption of various IT technologies and firm performance for a sample of publicly traded companies. Dr. Doms also is examining models of IT investment at the national level; this involves testing the various hypotheses surrounding the 1990s surge and 2001 sharp drop in IT investment.

Among Dr. Doms' published work are "How Fast Do Personal Computers Depreciate? Concepts and New Estimates" (with Wendy E. Dunn, Stephen D. Oliner, and Daniel E. Sichel. FRBSF Working Paper 2003-20. November); "IT Investment and Firm Performance in U.S. Retail Trade" (with Ron S. Jarmin, and Shawn D. Klimek. FRBSF Working Paper 2003-19 November); and "Understand-

---

ing Productivity: Lessons from Longitudinal Microdata" (with Erik Bartelsman, *Journal of Economic Literature*, September, pp. 569–594, September 2000) as well as other articles appearing in journals such as *Review of Economic Dynamics, Economics of Innovation and New Technology, Quarterly Journal of Economics, Economic Inquiry*, and *International Journal of Industrial Organization*.

Mark Doms received a B.A. from the University of Maryland and a Ph.D. in economics from the University of Wisconsin.

CHARLES H. FERGUSON

Charles H. Ferguson is a nonresident senior fellow in Economic Studies at the Brookings Institution and an independent computer consultant. He is author of The Broadband Problem: Anatomy of a Market Failure and a Policy Dilemma (Brookings Institution Press, 2004), *High Stakes, No Prisoners: A Winner's Talk of Greed and Glory in the Internet Wars* (Times Books, 1999) and coauthor with Charles R. Morris of *Computer Wars: The Fall of IBM and the Future of Global Technology* (Random House, 1993). He founded and served as CEO of Vermeer Technologies, the company responsible for developing FrontPage. He received a Ph.D. in political science from MIT.

KENNETH FLAMM

Kenneth Flamm is Professor and Dean Rusk Chair in International Affairs at the LBJ School of Public Affairs at UT–Austin. He is a 1973 honors graduate of Stanford University and received a Ph.D. in economics from M.I.T. in 1979. From 1993 to 1995, Dr. Flamm served as Principal Deputy Assistant Secretary of Defense for Economic Security and Special Assistant to the Deputy Secretary of Defense for Dual Use Technology Policy. Prior to and after his service at the Defense Department, he spent eleven years as a Senior Fellow in the Foreign Policy Studies Program at Brookings. Dr. Flamm has been a professor of economics at the Instituto Tecnológico A. de México in Mexico City, the University of Massachusetts, and George Washington University.

Dr. Flamm currently directs the LBJ School's Technology and Public Policy Program, and directs externally funded research projects on "Internet Use in Developing and Industrializing Countries." "The Economics of Fair Use," and "Determinants of Internet Use in U.S. Households," and has recently initiated a new project on "Exploring the Digital Divide: Regional Differences in Patterns of Internet Use in the U.S." He continues to work with semiconductor industry research consortium International SEMATECH, and is building a return-on-investment-based prototype to add economic logic to SEMATECH's industry investment model. He also is a member of the National Academy of Science's Panel on The Future of Supercomputing, and its Committee on Measuring and Sustaining the New Economy. He has served as member and Chair of the NATO

Science Committee's Panel for Science and Technology Policy and Organization, and as a member of the Federal Networking Council Advisory Committee, the OECD's Expert Working Party on High Performance Computers and Communications, and various advisory committees and study groups of the National Science Foundation, the Council on Foreign Relations, the Defense Science Board, and the U.S. Congress' Office of Technology Assessment, and as a consultant to government agencies, international organizations, and private corporations.

Dr. Flamm is the author of numerous articles and books on the economic impacts of technological innovation in a variety of high technology industries. Among the latter are *Mismanaged Trade? Strategic Policy and the Semiconductor Industry* (1996), *Changing the Rules: Technological Change, International Competition, and Regulation in Communications* (ed., with Robert Crandell, 1989), *Creating the Computer* (1988), and *Targeting the Computer* (1987). Recent work by Flamm has focused on measurement of the economic impact of the semiconductor industry on the U.S. economy, analyzing the economic determinants of Internet use by households, and assessing the economic impacts of Internet use in key applications.

LISA A. HOOK

As president of AOL Broadband, Premium & Developer Services, Lisa A. Hook oversaw AOL's drive to offer the premier broadband experience to AOL members. AOL for Broadband develops, markets, and operates AOL's high-speed line of business with more than 2.5 million subscribers, as of September 30, 2003.

In addition, Ms. Hook led the Premium Services organization, which develops, launches, and operates new subscription services. Working in tandem with AOL's marketing organization, the Premium Services group works across AOL to define and quickly deploy services that bring even greater value to members' online experience.

Ms. Hook's responsibilities also included oversight of AOL's Platform Services initiative that develops the next-generation platform strategy needed to launch new technologies in a scalable manner, including concerns such as transactions and authentication.

Formerly, Ms. Hook served as president of AOL Anywhere. In this role, she directed strategy and oversaw daily operations for the company's fast-growing Anywhere division and its mobile and voice services and non-PC devices. She also was responsible for new initiatives and partnerships to bring AOL's hallmark convenience and ease-of-use to online consumers beyond the PC worldwide.

A widely respected veteran of the telecommunications and media business, Ms. Hook joined AOL in 2000 as senior vice president of AOL Mobile and served as senior vice president and chief operating officer of that division before she was named president of AOL Anywhere.

Prior to joining the company, Ms. Hook was a partner at Brera Capital Partners LLC, a private equity firm focused on investing in media and telecommunications. Hook also has served as executive vice president and chief operating officer of Time Warner Telecommunications and later was vice president of Time Warner, Inc. managing various telecom-related transactions and operating matters. At Time Warner, she established the company's cellular and paging resale business and built its first successful retail outlets for cellular, paging, and cable services.

Ms. Hook initially joined Time Warner in 1989 as special advisor to the vice chairman of Time Warner Inc. In this position, she developed and oversaw international joint ventures including the acquisition of cable systems and the launch of theatre and cable services in Europe.

Earlier in her career, Ms. Hook served as legal advisor to the chairman of the Federal Communications Commission. Before that, she served as senior attorney at Viacom International Inc., where she oversaw the legal department of Viacom's cable division. Ms. Hook was also an associate with the law firm of Hogan & Hartson in Washington, D.C.

Ms. Hook is a director of National Geographic Ventures and a member of the Board of Trustees of the National Public Radio Foundation.

Ms. Hook is a graduate of Duke University and the Dickinson School of Law.

DAVID S. ISENBERG

In 1997, David S. Isenberg wrote an essay entitled, *The Rise of the Stupid Network: Why the Intelligent Network was a Good Idea Once but isn't Anymore.* In it, Isenberg (then a Distinguished Member of Technical Staff at AT&T Laboratories) examined the technological bases of the existing telecom business model, laid out how the communications business would be changed by new technologies, foresaw today's cataclysms, and imagined tomorrow's new network.

Tom Evslin, a senior AT&T executive at that time, told *The Wall Street Journal* that *The Rise of the Stupid Network*, "was like a glass of cold water in the face" of AT&T's leaders. *The Wall Street Journal* called the essay "scathing . . . startling," and said, "it may soon assume cult status among the tech mavens that roam the World Wide Web." Communications Week International said that the essay "challenged the most sacred assumptions of the telecom world." The Gilder Technology Report said it was "a stirring call." Inevitably, the essay found wider acceptance outside of AT&T than within it. So in 1998, Isenberg left AT&T to found isen.com, inc. to help telecommunications companies understand the business implications of the newly emerging communications infrastructure.

Dr. Isenberg's public delivery of the Stupid Network message is passionate and personal. He has spoken to over 100 audiences on three continents. For example, he has spoken numerous times at George Gilder's Telecosm, at Jeff

Pulver's Voice on the Net, at Kevin Werbach's SuperNova, at John McQuillan's Next Generation Networks, at the Canadian Advanced Network Research (CANARIE) annual meeting, at Merrill Lynch and Chase Bank telecom investor meetings, at the International Institute of Communications, at the Asia Pacific Regional Internet Conference (APRICOT), at the Optoelectronics Industry Development Association (OIDA) annual conference, at the Fiber to the Home Council's first annual meeting, and at numerous private management, customer, investor, and technology events.

Dr. Isenberg has been cited and quoted in *The New York Times, The Wall Street Journal, USA Today, Forbes, Fortune, Wired, Business 2.0, Communications Week International, Network World, Release 1.0, Gilder Technology Report, TheStreet.com, Nikkei Communications,* and numerous other publications. His story appears in at least half a dozen business books, including *Telecosm* by George Gilder, *The New Pioneers* by Tom Petzinger, and *The Future of Ideas* by Lawrence Lessig.

Dr. Isenberg has written articles for *Fortune, USA Today, IEEE Spectrum,* MSNBC, *Communications Week International, Light Reading, Business 2.0, America's Network, VON Magazine,* and *ACM Networker.* Isenberg advises a number of new telecommunications companies and their investors. He serves as a member of TechBrains (the Merrill Lynch technology strategy advisory board). He sits on advisory boards of CallWave, LaunchCyte, Broadband Physics, Terabeam, and YottaYotta.

Dr. Isenberg is a fellow of Glocom, the Institute for Global Communications of the International University of Japan. He is a founding advisor of the World Technology Network. He was a judge of the World Communications Awards in 1999 and 2001.

In his 12-year career at AT&T (1985–1998), Dr. Isenberg was a distinguished member of Technical Staff with AT&T Labs Research, the part of Bell Labs that stayed with AT&T after the 1996 "trivestiture." Before that, he held AT&T Bell Labs technical positions in Consumer Long Distance, in Network Services, and in the PBX business unit. Before AT&T, Dr. Isenberg was employed by Mattel and Verbex, and did consulting work in voice processing for Milton Bradley, National Semiconductor, GTE Labs, and others. David Isenberg holds a Ph.D. in biology from the California Institute of Technology (1977) but also learned much science growing up in Woods Hole, Massachusetts. His upbringing centered around two principles: (1) Research is useful, and (2) If you are going to fish, use a big hook.

JEFFREY M. JAFFE

Jeffrey M. Jaffe is president of Bell Labs Research and Advanced Technologies for Lucent Technologies. Bell Labs, the company's global research and development arm, consists of approximately 10,000 employees in 10 countries.

As president of Bell Labs Research, Dr. Jaffe supports basic research to advance science and technology in areas of importance to Lucent. Advanced Technologies works with Lucent's business units in the commercial development and deployment of new technologies.

Prior to joining Lucent in 2000, Dr. Jaffe held a variety of executive research positions with International Business Machines (IBM) in a 20-year career, which included general manager of SecureWay Software and Corporate Vice President of Technology.

Dr. Jaffe is a fellow of the IEEE and the Association of Computing Machinery. The United States government has consulted with Dr. Jaffe on numerous policy initiatives. In 1997, President Clinton appointed Dr. Jaffe to the Advisory Committee for the President's Commission for Critical Infrastructure Protection. Dr. Jaffe has chaired the Chief Technology Officer Group of the Computer Systems Policy Project (CSPP), which consists of a dozen of the top computer and telecommunications companies, and has served on The National Research Council's Computer Science & Telecommunications Board.

Dr. Jaffe earned his B.S. in mathematics, a M.S. in electrical engineering, and a Ph.D. in computer science, from the Massachusetts Institute of Technology.

DALE W. JORGENSON

Dale W. Jorgenson is the Samuel W. Morris University Professor at Harvard University. He received a B.A. in economics from Reed College in Portland, Oregon, in 1955 and a Ph.D. in economics from Harvard in 1959. After teaching at the University of California–Berkeley, he joined the Harvard faculty in 1969 and was appointed the Frederic Eaton Abbe Professor of Economics in 1980. He has directed the Program on Technology and Economic Policy at the Kennedy School of Government since 1984 and served as chairman of the Department of Economics from 1994 to 1997.

Dr. Jorgenson has been honored with membership in the American Philosophical Society (1998), the Royal Swedish Academy of Sciences (1989), the U.S. National Academy of Sciences (1978), and the American Academy of Arts and Sciences (1969). He was elected to fellowship in the American Association for the Advancement of Science (1982), the American Statistical Association (1965), and the Econometric Society (1964). He has been awarded honorary doctorates by Uppsala University (1991), the University of Oslo (1991), Keio University (2003), and the University of Mannheim (2004).

Dr. Jorgenson served as president of the American Economic Association in 2000 and was named a distinguished fellow of the Association in 2001. He was a founding member of the Board on Science, Technology, and Economic Policy of the National Research Council in 1991 and has served as chairman of the Board since 1998. He also served as chairman of Section 54, Economic Sciences, of the

National Academy of Sciences from 2000 to 2003 and was president of the Econometric Society in 1987.

Dr. Jorgenson received the prestigious John Bates Clark Medal of the American Economic Association in 1971. This Medal is awarded every two years to an economist under forty for excellence in economic research. The citation for this award reads in part:

> *Dale Jorgenson has left his mark with great distinction on pure economic theory (with, for example, his work on the growth of a dual economy); and equally on statistical method (with, for example, his development of estimation methods for rational distributed lags). But he is preeminently a master of the territory between economics and statistics, where both have to be applied to the study of concrete problems. His prolonged exploration of the determinants of investment spending, whatever its ultimate lessons, will certainly long stand as one of the finest examples in the marriage of theory and practice in economics.*

Dr. Jorgenson has conducted groundbreaking research on information technology and economic growth, energy and the environment, tax policy and investment behavior, and applied econometrics. He is the author of 232 articles in economics and the author and editor of twenty-four books. His collected papers have been published in ten volumes by The MIT Press, beginning in 1995. His most recent book, *Economic Growth in the Information Age*, was published by The MIT Press in 2002 and represents the first major effort to quantify the impact of information technology on the U.S. economy. Another recent MIT Press volume, *Lifting the Burden: Tax Reform, the Cost of Capital, and U.S. Economic Growth*, co-authored with Kun-Young Yun in 2001, proposes a new approach to capital income taxation, dubbed "A Smarter Type of Tax" by the *Financial Times*.

Forty-three economists have collaborated with Dr. Jorgenson on published research. Many of Dr. Jorgenson's books and papers have been co-authored with students in economics at Berkeley and Harvard. Among his former students are professors at leading academic institutions in the United States and abroad and several occupy endowed chairs. The MIT Press published *Econometrics and the Cost of Capital*, edited by Lawrence J. Lau, in 2000. This contains essays in honor of Dr. Jorgenson presented at a conference at Harvard by thirteen of his former students. It also contains his biography, a list of his publications, and a list of his sixty-four Ph.D. thesis advisees at Berkeley and Harvard.

MIKE LAJOIE

Mike LaJoie is chief technology officer of Time Warner Cable. Prior to his appointment to CTO, Mr. LaJoie had been serving as the company's executive vice president of Advanced Technology since August 2002.

Over the last several years, Mr. LaJoie has lead the development and deployment of Time Warner Cable's extremely successful advanced digital products including video on demand, high definition television and digital video recorders. As chief technology officer, Mr. LaJoie guides technology development across all product offerings at the company. He also charts the course for the continuing evolution of the company's digital platform and technological infrastructure. Mr. LaJoie continues to build on his responsibility for new technology development, set-top advances and industry standards activities, such as OCAP and DOCSIS, while driving to keep Time Warner Cable in its position as technology leader within the industry.

Mr. LaJoie served as vice president of Corporate Development from 1998 through 2002, where he oversaw the development of VOD software and set-top boxes and other major launches of new services and products. Mr. LaJoie has been involved in many development projects over his 16 years working with the company, including its Multi Media initiative, The Full Service Network, Road Runner, Pegasus Digital Television platform, and the company's early work in IP telephony.

Prior to joining Time Warner Cable, Mr. LaJoie was an independent software developer and designed and installed network systems. Earlier in his career he was a NASDAQ Broker/Dealer and a Series 7 Registered Securities Representative.

DAVID LIPPKE

David Lippke, president of HighSpeed America, has over two decades of intensive, industry-leading experience in technology development and management. Mr. Lippke is particularly well known for his openness, the strength of his inter-organizational relationships, and his desire to understand others' perspectives.

Mr. Lippke joined America Online (AOL) in 1993 where he led the development of the company's core infrastructures, scaling mechanisms, and key applications for nine years. Early in his AOL career, Mr. Lippke developed AOL's scalable architecture including the architecture and implementation of AOL Instant Messenger™, a high-performance messaging fabric architected in 1996 to support the two orders of magnitude growth represented by the nine million simultaneous user load experienced now.

Mr. Lippke most recently served as AOL's senior vice president for Systems Infrastructure leading a nationwide engineering organization of some 9,000 employees with primary responsibility for the company's host-based products; systems and connectivity infrastructures; as well as its advertising, publishing, and content-tracking systems.

LOUIS MAMAKOS

Louis Mamakos, chief technology officer, oversees all technology functions at Vonage, which includes new product and services development, supervision of all research projects and integration of all technology-based activities into Vonage's corporate strategy.

Mr. Mamakos has more than 20 years of experience in Internet technical engineering and architecture for large scale, commercial IP backbone networks. Most recently, Mr. Mamakos served as a fellow for Hyperchip, Inc., a start-up that built scalable, high-performance core routers. Prior to Hyperchip, Mr. Mamakos held various engineering and architecture positions during his eight years at UUNET Technologies, now known as MCI. Prior to UUNET Technologies, Mr. Mamakos spent nearly twelve years as Assistant Manager for Network Infrastructure at the University of Maryland, College Park.

Mr. Mamakos holds a B.S. in computer science from University of Maryland—College Park.

STEVEN J. METALITZ

Steven J. Metalitz is a Partner in the Washington, D.C. law firm of Smith & Metalitz, LLP. He specializes in intellectual property, privacy, e-commerce and information law. He provides legal counseling and policy advocacy, primarily for clients in the publishing, recording, motion picture, software and database industries, and for e-commerce companies.

For the past decade, Mr. Metalitz has represented the main coalitions of the copyright industry sector on key public policy issues. For example, as counsel to the Creative Incentive Coalition, Mr. Metalitz was closely involved in the drafting and enactment of the Digital Millennium Copyright Act of 1998, and since then he has represented a copyright industry coalition on DMCA implementation matters. He also serves as senior vice president of the International Intellectual Property Alliance® (IIPA®), the coalition of copyright industry trade associations working for stronger copyright protection and enforcement around the world, including ratification and implementation of the WIPO Internet treaties. He has been counsel to the Copyright Coalition on Domain Names (CCDN) since its establishment in 1999, and has been an officer of the Intellectual Property Constituency of the Internet Corporation for Assigned Names and Numbers (ICANN) since its inception, including two terms as president.

From 1989–1994, Mr. Metalitz was vice president and general counsel of the Information Industry Association, directing the trade association's government relations program and developing and advocating its policy positions in copyright, telecommunications, privacy, government information policy, and other areas. From 1982–1989, he held several senior staff positions with the U.S. Senate Judiciary Committee, including chief nominations counsel, and chief counsel and staff director of its Subcommittee on Patents, Copyright and Trademarks. He also served as legislative director to Senator Charles McCurdy Mathias, Jr. (R-MD). Before his government service, Mr. Metalitz practiced law in Charleston, South Carolina. Mr. Metalitz is a member of the bar in the District of Columbia and South Carolina (inactive). He has taught copyright law as professorial lecturer in law at the George Washington University Law School in Washington, D.C., and has published widely on copyright and Internet law topics. He is a Phi Beta Kappa graduate of the University of Chicago (B.A. 1972) and earned his law degree at Georgetown University Law Center (J.D. 1977).

CHERRY A. MURRAY

Cherry A. Murray, Research Strategy, Wireless and Physical Sciences Research senior vice president at Bell Laboratories, Lucent Technologies, is a physicist recognized for her work in surface physics, light scattering, and complex fluids. She is best known for her imaging work in phase transitions of colloidal systems. Dr. Murray was born in 1952 in Ft. Riley Kansas into an Army and then Foreign Service family, and spent her childhood traveling around the world, moving on the average of once per year. After receiving a B.S. and Ph.D in physics from the Massachusetts Institute of Technology, she was hired into Bell Labs as a Member of Technical Staff in 1978. Dr. Murray became a distinguished member of Technical Staff at Bell Labs in 1985. She has numerous publications and two patents.

At Bell Labs, Dr. Murray was promoted to department head of the Low Temperature Physics Department in 1987 and served as department head of the Condensed Matter Physics, and then Semiconductor Physics Departments until 1997, when she was promoted to director, Physical Research Lab. She is proud of managing the 40Gb/s electronics group and the invention and development of the optical fabric for the first all-optical crossconnect for telecommunications networks, Lucent's Wavestar LambdaRouter. She was promoted to Physical Sciences senior vice president in April 2000, and assumed her present responsibilities in October 2001. In this position, Dr. Murray has responsibility for the strategy of all Bell Labs Research and also Bell Labs Research Business Development. She also manages the Wireless and Physical Research Labs, and is responsible for the relationship of Bell Labs Research with Lucent's largest business unit, Mobility Solutions.

Dr. Murray is a member of the National Academy of Sciences, the National Academy of Engineering and the American Academy of Art and Sciences. She is a fellow of the American Physical Society and the American Association for the Advancement of Science and a member of the American Chemical Society, the Optical Society of America, the Materials Research Society, and Sigma Xi. She won the APS Maria Goeppert-Mayer Award in 1989. She sits on numerous advisory committees and boards, including the National Sciences Resource Center, dedicated to the propagation of inquiry-based science education. She is currently a General Councilor of the American Physical Society, a councilor of the National Academy of Sciences, the National Research Council, and the University of Chicago Board of Governors of Argonne National Laboratory. She also serves on the Basic Energy Sciences Advisory Committee for the Department of Energy.

MARK B. MYERS

Mark B. Myers is visiting executive professor in the Management Department at the Wharton Business School, the University of Pennsylvania. His research interests include identifying emerging markets and technologies to enable growth in new and existing companies with special emphases on technology identification and selection, product development and technology competencies. Dr. Myers serves on the Science, Technology and Economic Policy Board of the National Research Council and currently co-chairs with Richard Levin, the President of Yale, the National Research Council's study of "Intellectual Property in the Knowledge Based Economy."

Dr. Myers retired from the Xerox Corporation at the beginning of 2000, after a 36-year career in its research and development organizations. Dr. Myers was the Senior Vice President in charge of corporate research, advanced development, systems architecture and corporate engineering from 1992 to 2000. His responsibilities included the corporate research centers, PARC in Palo Alto, California; Webster Center for Research & Technology near Rochester, New York; Xerox Research Centre of Canada, Mississauga, Ontario; and the Xerox Research Centre of Europe in Cambridge, UK, and Grenoble, France. During this period was a member of the senior management committee in charge of the strategic direction setting of the company.

Dr. Myers is chairman of the board of trustees of Earlham College and has held visiting faculty positions at the University of Rochester and at Stanford University. He holds a bachelor's degree from Earlham College and a doctorate from Pennsylvania State University.

MICHAEL R. NELSON

As Director of Internet Technology and Strategy at IBM, Michael R. Nelson manages a team helping define and implement IBM's Next Generation Internet strategy (NGi). His group works with university researchers on NGi technology and shaping standards for the NGi. He is also responsible for organizing IBM's involvement in the Internet2 research consortium. He chaired the Internet Society's annual INET2002 meeting and was recently selected as the Society's Vice President for Public Policy.

Prior to joining IBM in July 1998, Dr. Nelson was director for technology policy at the Federal Communications Commission (FCC), an independent United States government agency that is charged with regulating interstate and international communications by radio, television, wire, satellite and cable. There he helped craft policies to foster electronic commerce, spur development and deployment of new technologies, and improve the reliability and security of the nation's telecommunications networks.

Before joining the FCC in January 1997, Dr. Nelson was special assistant for information technology at the White House Office of Science and Technology Policy. Here he worked with Vice President Gore and the President's Science Advisor on issues relating to the Global Information Infrastructure, including telecommunications policy, information technology, encryption, electronic commerce, and information policy.

From 1988 to 1993 Dr. Nelson served as a professional staff member for the Senate's Subcommittee on Science, Technology, and Space, chaired by then-Senator Gore. He was the lead Senate staffer for the High-Performance Computing Act.

Michael Nelson has a B.S. in geology from Caltech, and a Ph.D. in geophysics from MIT.

WILLIAM J. RADUCHEL

William J. Raduchel is the chairman and chief executive officer of Ruckus Network bringing a broad range of business experience in the computing, Internet and media industries. Before joining Ruckus Network, Dr. Raduchel was executive vice president and chief technology officer of AOL Time Warner, Inc.

Prior to AOL, Dr. Raduchel served as chief strategy officer and an executive committee member for Sun Microsystems, Inc. In his eleven years at Sun, Dr. Raduchel also held positions as chief information officer, chief financial officer, acting vice president of human resources and vice president of corporate planning and development and oversaw relationships with major Japanese partners. In addition, he has held senior executive roles at Xerox Corporation and McGraw-Hill, Inc.

Dr. Raduchel currently serves as a director of Chordiant Software, In2Books, PanelLink Cinema Partners PLC and as an adviser to its parent company, Silicon Image. Additionally, he is an adviser to Myriad International Holdings, Hyperspace Communications and Wild Tangent. Dr. Raduchel is a member of the National Advisory Board for the Salvation Army, the National Academy Committee on Internet Navigation and Domain Name Services and the Board on Science, Technology and Economic Policy of the National Academy of Sciences.

Named "CTO of the Year" in 2001 by Infoworld magazine, Dr. Raduchel was a past professor of economics at Harvard University and holds several issued and pending patents. After attending Michigan Technological University, which gave him an honorary doctorate in 2002, Dr. Raduchel received his undergraduate degree in economics from Michigan State University and earned his A.M. and Ph.D. degrees in economics at Harvard. In both the fall and spring of 2003 he was the Castle Lecturer on Computer Science at the U.S. Military Academy at West Point.

ANDREW SCHUON

Andrew Schuon is the president of International Music Feed. Formerly he was president of programming for Infinity Broadcasting, where he was responsible for group-wide programming for 183 radio stations, and president and chief executive officer of Pressplay, where he oversaw all aspects of Pressplay's operations, including the launch of the online subscription service, the management of its technical operations and the overall branding and development of the service.

Prior to joining Pressplay, Mr. Schuon was president and chief operating officer of Jimmy and Doug's Farmclub.com where he was responsible for overseeing all aspects of the company's record label operations, online activities, and television programming since its launch in January 2000. Previously, Mr. Schuon was executive vice president and general manager of Warner Brothers Records, with responsibility for all creative and administrative issues including promotion, marketing, artist relations, advertising, art, sales, and production. Before his post at Warner Brothers, Mr. Schuon spent several years at MTV, Music Television, culminating in his title as executive vice president of programming. Mr. Schuon is credited with engineering the station's evolution from "video jukebox" to a fully realized "youth culture" network. He was the executive producer of the MTV Video Music Awards and the MTV Movie Awards, and created and developed such programming as "Alternative Nation," "MTV Live (now TRL)," "MTV Jams," and "The MTV Beach House." Mr. Schuon also served as executive vice president of programming at VH-1 where he supervised the channel's successful re-launch.

PETER A. TENHULA

On April 7, 2003, Peter A. Tenhula was named acting deputy bureau chief of the FCC's Wireless Telecommunications Bureau. In this position, Mr. Tenhula oversees the Bureau's Mobility Division and its Auctions and Spectrum Access Division.

Mr. Tenhula also serves as director of the FCC's Spectrum Policy Task Force where he is leading the next phase of the Task Force's mission, including the coordination of spectrum policy activities within the FCC, with Congress and with the administration. He also serves on the FCC's Homeland Security Policy Council.

Before taking on his current duties, Mr. Tenhula served as senior legal advisor to Chairman Michael K. Powell. He advised Chairman Powell on various issues including matters related to wireless telecommunications, spectrum policy, international communications, and national security/emergency preparedness. Mr. Tenhula joined then-Commissioner Powell's staff as a legal advisor in 1997. A 13-year FCC veteran, Mr. Tenhula started his career at the FCC in 1990 as a staff attorney in the Video Services Division of the Mass Media Bureau. From 1991–1995, he worked in the Administrative Law Division of the FCC's Office of General Counsel, and from 1995–1997 he served as special counsel to the FCC's General Counsel.

Prior to joining the Commission, Mr. Tenhula served as a legal intern with U.S. Representative Michael G. Oxley and the National Association of Broadcasters.

Mr. Tenhula received a B.A. degree in telecommunications from Indiana University—Bloomington, and a law degree from Washington University in St. Louis, Missouri. He is a member of the Missouri Bar and the Federal Communications Bar Association.

H. BRIAN THOMPSON

Brian Thompson is the chairman and founder of iTown Communications. As a veteran senior executive of the telecommunication industry, Mr. Thompson has been instrumental in impacting the rise of competitive telecommunications both in the United States and abroad. Mr. Thompson continues to head his own private equity investment and advisory firm, Universal Telecommunications, Inc. in Vienna, Virginia, focused on both start-up companies and consolidations taking place in the information/telecommunications industries.

Mr. Thompson currently serves as Chairman, Comsat International (CI), one of the largest independent telecommunications operators serving all of Latin America. He was previously Chairman and chief executive officer of Global TeleSystems Group, Inc. from March 1999 through September of 2000. He served as chairman and CEO of LCI International, leading a turnaround of the company and developing it into one of the fastest growing telecommunications companies in the United States. Subsequent to the merger of LCI with Qwest Communications International Inc. in June 1998, he became vice chairman of the Board for Qwest until his resignation.

Mr. Thompson was also the executive vice president of MCI Communications Corporation during its formative years as a long distance service company from 1981 to 1990 with responsibility for the company's eight operating divisions, including MCI International.

Mr. Thompson currently serves as a member of the board of directors of Bell Canada International Inc., ArrayComm, Inc., Axcelis Technologies, Inc., Sonus Technologies, and United Auto Group. He also serves as the U.S. co-chairman of the Global Information Infrastructure Commission, a multinational organization charting the role of the private sector in the developing global information and telecommunications infrastructure. Additionally, he is a member of the Irish Prime Minister's Ireland-America Economic Advisory Board.

MARK A. WEGLEITNER

Mark A. Wegleitner is senior vice president, technology, and chief technology officer (CTO) for Verizon Communications. He is responsible for technology assessment, network architecture, technology planning, platform development, and laboratory infrastructure for the wireline communications business. In addition, he oversees a group providing technology solutions for government and commercial customers: Federal Network Systems. In his current role, he and his organization support all business units in the management of technology matters.

Prior to his current assignment, Mr. Wegleitner served as vice president, Technology & Engineering, at Bell Atlantic Network Services, where he was responsible for all technology and engineering functions. Prior to that, he was CTO at Bell Atlantic Network Services.

Since joining Bell Atlantic, Mr. Wegleitner has also held a variety of other management positions in strategic planning, network architecture, technology development, information systems, research and development, broadband implementation, and new services technology.

Mr. Wegleitner began his career in 1972 with Bell Telephone Laboratories in local switching systems development. In 1979, he joined the exchange switching systems design organization at AT&T General Departments, where he had responsibility for the introduction of new features and services on local switching systems. In 1983, he held a brief assignment with Bell Laboratories in local switching systems engineering before transferring to Bell Atlantic.

Mr. Wegleitner received a B.A. in mathematics from St. John's University, and an M.S. in electrical engineering and computer science from the University of California at Berkeley.

Appendix B

Participants List*

Jeffrey Alexander
Washington CORE

Caterina Au
George Mason University

Phil Auerswald
George Mason University

Stephanie Baroni
SETECS

Eugenie Barton
Federal Communications Commission

Kim Bayard
Federal Reserve Board

Bill Belt
Telecommunications Industry
  Association

Richard Bissell
The National Academies

Craig Burkhardt
Department of Commerce

Dave Byrne
Federal Reserve Board

Kenneth Carter
Federal Communications Commission

Lynn Chapman
George Mason University

Speakers in italics.

Barbara Cherry
Federal Communications Commission

McAlister Clabaugh
The National Academies

Denise Coca
Federal Communications Commission

Eileen Collins
Rutgers Center for Women and Work

Carol Corrado
Federal Reserve Board

Mark Crawford
Department of Commerce

Santanu Das
Office of Naval Research

Cynthia de Lorenzi
PatriotNet

David Dierksheide
The National Academies

Kyle Dixon
Progress and Freedom Foundation

Mark E. Doms
Federal Reserve Bank of San Francisco

Tom Donahue

Jesus Dumagan
Department of Commerce

Wendy Dunn
Federal Reserve Board

Charles H. Ferguson
The Brookings Institution

Jeffrey Fillar
National Communications System

Kevin Finneran
The National Academies

Kenneth Flamm
University of Texas at Austin

Barbara Fraumeni
Department of Commerce

John Gardenier

Turkan Gardenier
Pragmatica

Robin Gaster
North Atlantic Research

Behzad Ghaffari
Federal Communications Commission

Peter Gottstein
Embassy of Germany

Miguel Green
George Mason University

Henry Heilbrunn
Interactive Directions

Miriam Heller
National Science Foundation

Jay Hellman

Robert Hershey

Michael Holdway
Department of Labor

Lisa A. Hook
AOL Broadband (retired)

John Horrigan
Pew Internet Project

Richard Hovey
Federal Communications Commission

Tim Hughes
House Science Committee

Dan Hurley
Department of Commerce

David S. Isenberg
Isen.com

Ken Jacobson
National Academies

Jeffrey M. Jaffe
Lucent Technologies

Dale W. Jorgenson
Harvard University

Magnus Karlsson
Embassy of Sweden

Derek Khlopin
Telecommunications Industry
    Association

Jean-Phillipe Lagrange
Embassy of France

Mike LaJoie
Time Warner Cable

Emmanuel Le Perru
Embassy of France

Alfred Lee
Department of Commerce

Richard Lempert
National Science Foundation

Tom Lenard
Progress and Freedom Foundation

Paul Lengermann
Federal Reserve Board

Nanette Levinson
American University

David Lippke
HighSpeed America

Nancy Lutz
National Science Foundation

Neil MacDonald
Federal Technology Watch

Sira Maliphol
Embassy of the United Kingdom

Louis Mamakos
Vonage

Goran Marklund
VINNOVA

Hugh McElrath
Office of Naval Intelligence

Pamela Megna
Federal Communications Commission

Bart Meroney
Department of Commerce

Stephen J. Metalitz
Smith & Metalitz

Clark Misul
DETECON

Sabrina Montes
Department of Commerce

Cherry A. Murray
Lucent Technologies

Nathan Musick
Congressional Budget Office

Mark B. Myers
The Wharton School
University of Pennsylvania

Michael R. Nelson
International Business Machines

Kent Nilsson
Federal Communications Commission

Sue Okubo
Department of Commerce

Luciano Parodi
Embassy of Chile

David Peyton
National Association of Manufacturers

Peter Pitsch
Intel Corporation

Lorenzo Pupillo
The World Bank

William J. Raduchel
Ruckus Network

Stan Riveles
State Department

Dorothy Robyn
The Brattle Group

Andrew Schuon
International Music Feed

Sujai Shivakumar
The National Academies

Robert Sienkiewicz
National Institute of Standards and
 Technology

Meredith Singer
Telecommunications Industry
 Association

Christina Siroskey
George Washington University

Jim Snider
New America Foundation

Anna Snow
European Commission

Thomas Spavins
Federal Communications Commission

Francessca Spencer
Oracle

Kenneth Spratt
Embassy of Ireland

Diane Steinour
Department of Commerce

David Su
National Institute of Standards and
    Technology

Peter A. Tenhula
Federal Communications Commission

H. Brian Thompson
iTown Communications

Hannele Tikkanen
Embassy of Finland

Richard Van Atta
Institute for Defense Analyses

Venu Veeravalli
National Science Foundation

David Wasshausen
Department of Commerce

Joseph Watson
Department of Commerce

Philip Webre
Congressional Budget Office

Mark A. Wegleitner
Verizon

Josh Weiner
Ruckus Network

Charles W. Wessner
The National Academies

John Wohlstetter
Discovery Institute

Appendix C

Selected Bibliography

Abel, Jaison R., Ernst R. Berndt, Alan G. White. 2003. *Price Indexes for Microsoft's Personal Computer Software Products.* NBER Working Paper 9966. Cambridge, MA: National Bureau of Economic Research.

Abramovitz, Moses, and Paul David. 1999. "American Macroeconomic Growth in the Era of Knowledge-Based Progress: The Long Run Perspective." In *Cambridge Economic History of the United States.* Robert E. Gallman and Stanley I. Engerman, eds. Cambridge, UK: Cambridge University Press.

Advanced Technology Program. 2002. "Benefits and Costs of ATP Investments in Component-Based Software." NIST GCR 02-834. Gaithersburg, MD: U.S. Department of Commerce.

Aizcorbe, Ana. 2005. "Moore's Law, Competition, and Intel's Productivity in the Mid-1990s." BEA Working Paper WP2005-8. Washington, D.C.: Bureau of Economic Research.

Aizcorbe, Ana. 2005. "Why are Semiconductor Price Indexes Falling so Fast? Industry Estimates and Implications for Productivity Measurement." BEA Working Paper WP2005-7. Washington, D.C.: Bureau of Economic Research.

Aizcorbe, Ana, Kenneth Flamm, and Anjum Khurshid. 2002. "The Role of Semiconductor Inputs in IT Hardware Price Decline: Computers vs. Communications." Federal Reserve Board Finance and Economics Series Discussion Paper 2002-37. Washington, D.C.: Board of Governors of the Federal Reserve System.

Aizcorbe, Ana, Stephen D. Oliner, and Daniel E. Sichel. 2003. "Trends in Semiconductor Prices: Breaks and Explanations." Washington D.C.: Board of Governors of the Federal Reserve System.

Baily, Martin N. 2002. "The New Economy: Post Mortem or Second Wind?" *Journal of Economic Perspectives* 16(2):3–22.

Baily, Martin N. and Robert J. Gordon. 1998. "The Productivity Slowdown, Measurement Issues, and the Explosion of Computer Power." *Brookings Papers on Economic Activity* 2:347–420.

Baily, M. N. and R. Z. Lawrence. 2001. "Do We Have an E-conomy?" NBER Working Paper 8243. Cambridge, MA: National Bureau of Economic Research.

Bard, Yonathan and Charles H. Sauer. 1981. "IBM Contributions to Computer Performance Modeling." *IBM Journal of Research and Development* 25:562–570.

Barzyk, Fred. 1999. "Updating the Hedonic Equations for the Price of Computers." Statistics Canada Prices Division Working Paper. Ottawa, ON: Statistics Canada.

Bell, C. Gordon. 1986. "RISC: Back to the Future?" *Datamation* 32(June):96–108.

Benkard, C. Lanier. 2001. *A Dynamic Analysis of the Market for Wide Bodied Commercial Aircraft.* Stanford, CA: Graduate School of Business, Stanford University.

Berndt. Ernst R. and Jack E. Triplett, eds. 1990. *Fifty Years of Economic Measurement.* Chicago, IL: University of Chicago Press.

Berndt, Ernst R., and Zvi Griliches. 1993. "Price Indexes for Microcomputers: An Exploratory Study." In Murray F. Foss, Marilyn Manser, and Allan H. Young, eds. *Price Measurements and Their Uses.* Studies in Income and Wealth 57:63–93. Chicago: University of Chicago Press for the National Bureau of Economic Research.

Berndt, Ernst R., Zvi Griliches, and Neal Rappaport. 1995. "Econometric Estimates of Prices in Indexes for Personal Computers in the 1990s." *Journal of Econometrics* 68(1995):243–268.

Berndt, Ernst R. and Neal J. Rappaport. 2001. "Price and Quality of Desktop and Mobile Personal Computers: A Quarter-Century Historical Overview." *American Economic Review* 91(2):268–273.

Berndt, Ernst R. and Neal J. Rappaport. 2002. "Hedonics for Personal Computers: A Reexamination of Selected Econometric Issues." Unpublished Paper.

Bleha, Thomas, 2005. "Down to the Wire," *Foreign Affairs* 84(3).

Blinder, Alan, 1997. "The Speed Limit: Fact and Fancy in the Growth Debate." *The American Prospect* 8(34).

Bloch, Erich and Dom Galager. 1978. "Component Progress: It's Effect on High-Speed Computer Architecture and Machine Organization." *Computer* 11(April):64–75.

Bosworth, Barry P. and Jack E. Triplett. 2001. "What's new about the New Economy? IT, Economic Growth, and Productivity." *International Productivity Monitor* 2(Spring):19–30.

Bosworth, Barry P. and Jack E. Triplett. 2003. *Services Productivity in the United States: Griliches' Services Volume Revisited.* Washington, D.C.: The Brookings Institution.

Bourot, Laurent. 1997. "Indice de Prix des Micro-ordinateurs et des Imprimantes: Bilan d'une rénovation." Working Paper of the Institut National De La Statistique Et Des Etudes Economiques (INSEE). Paris, France.

Brainard, Lael and Robert E. Litan. 2004. *"Off-shoring" Service Jobs: Bane or Boon and What to Do?* Brookings Institution Policy Brief 132. Washington, D.C.: The Brookings Institution.

Bresnehan, Timothy, Eric Brynjolfsson, and Lorin M. Hitt. 2002. "Information Technology, Workplace Organization, and the Demand for Skilled Labor: Firm-level Evidence." *Quarterly Journal of Economics* 117(1):339–376.

Bromley, D. Alan. 1972. *Physics in Perspective.* National Academy of Sciences. Washington, D.C.: National Academy Press.

Brooks, Frederick. 1975. *The Mythical Man Month: essays on software engineering.* Reading, MA: Addison-Wesley Publishing Company.

Brynjolfsson, Eric and Lorin M. Hitt. 2000. "Beyond Computation: Information Technology, Organizational Transformation, and Business Performance." *Journal of Economic Perspectives* 14(4):23–48.

Brynjolfsson, Eric and Lorin M. Hitt. 2002. Computing Productivity: Firm-level Evidence." *Review of Economics and Statistics* 85(4):793–808.

Brynjolfsson, Eric and Brian Kahin, eds. 2000. *Understanding the Digital Economy.* Cambridge, MA: The MIT Press.

Bureau of Economic Analysis. 2001. "A Guide to the NIPAs." In *National Income and Product Accounts of the United States, 1929-97.* Washington, D.C.: U.S. Government Printing Office. Accessible at <http://www.bea.doc.gov/bea/an/nipaguid.pdf>.

Businessweek. 1997. "Deconstructing the Computer Industry." August 25.

Butler Group. 2001. "Is Clock Speed the Best Gauge for Processor Performance?" *Server World Magazine* September 2001. Accessed at <http://www.serverworldmagazine.com/opinionw/2001/09/06_clockspeed.shtml> on February 7, 2003.

Carr, Nicholas. 2005. "The End of Corporate Computing." *MIT Sloan Management Review* 46(3):67–73.

Cartwright, David W., Gerald F. Donahoe, and Robert P. Parker. 1985. "Improved Deflation of Computer in the Gross National Product of the United States." Bureau of Economic Analysis Working Paper 4. Washington, D.C.: U.S. Department of Commerce.

Cecchetti, Stephen G. 2002. "The New Economy and the Challenge for Macroeconomic Policy." Paper prepared for the conference, *The New Economy: What's new about it?* Texas A&M University. April 19.

Chandler, Alfred D. Jr. 2000. "The Information Age in Historical Perspective." In *A Nation Transformed by Information: How Information Has Shaped the United States from Colonial Times to the Present,* eds. Alfred D. Chandler and James W. Cortada. New York: Oxford University Press.

Choi, Soon-Yong and Andrew B. Whinston 2000. *The Internet Economy: Technology and Practice.* Austin, TX: SmartEcon Publishing.

Chow, Gregory C. 1967. "Technological Change and the Demand for Computers." *American Economic Review* 57(December):1117–1130.

Christiansen, Clayton. 1997. *The Innovator's Dilemma: When New Technologies Cause Great Firms to Fail.* Boston, MA: Harvard Business School Press.

Chwelos, Paul. 2003. "Approaches to Performance Measurement in Hedonic Analysis: Price Indexes for Laptop Computers in the 1990s." *Economics of Innovation and New Technology* 12(3):199–224.

Cohen Stephen S., and John Zysman. 1988. *Manufacturing Matters: The Myth of the Post-Industrial Economy.* New York: Basic Books.

Cohen, Wesley M. and John Walsh. 2002. "Public Research, Patents and Implications for Industrial R&D in the Drug, Biotechnology, Semiconductor and Computer Industries." In National Research Council. *Capitalizing on New Needs and New Opportunities: Government-Industry Partnerships in Biotechnology and Information Technologies.* Washington, D.C.: National Academy Press.

Cole, Rosanne, Y. C. Chen, Joan A. Barquin-Stolleman, Ellen Dulberger, Nurhan Helvacian, and James H. Hodge. 1986. "Quality-Adjusted Price Indexes for Computer Processors and Selected Peripheral Equipment." *Survey of Current Business* 66(1):41–50.

Colecchia, Alessandra and Schreyer, Paul. 2002. "ICT investment and economic growth in the 1990s: is the United States a unique case? A comparative study of nine OECD countries." *Review of Economic Dynamics* 5(2):408–442.

Congressional Budget Office. 2002. *The Role of Computer Technology in the Growth of Productivity.* Washington, D.C.: Congressional Budget Office.

Corrado, Carol A., John Haltiwanger, and Daniel Sichel, eds. 2005. *Measuring Capital in a New Economy.* Chicago, IL: University of Chicago Press.

Council on Competitiveness. 2004. *Innovate America, Thriving in a World of Challenge and Change,* Washington, D.C.: Council on Competitiveness.

Council of Economic Advisors. 2001. *Annual Report.* Washington, D.C.: U.S. Government Printing Office.

Council of Economic Advisors. 2002. *Annual Report.* Washington, D.C.: U.S. Government Printing Office.

Crandall, Robert C. and Kenneth Flamm, eds. 1989. *Changing the Rules: Technological Change, International Competition, and Regulation in Communications.* Washington, D.C.: The Brookings Institution.

Cunningham, Carl, Denis Fandel, Paul Landler, and Robert Wright. 2000. *Silicon Productivity Trends.* International SEMATECH Technology Transfer #00013875A-ENG.

Dalén, Jorgen. 1989. "Using Hedonic Regression for Computer Equipment in the Producer Price Index." R&D Report. Statistics Sweden. Research-Methods-Development. 1989:25.

David, Paul. 2000. "Understanding Digital Technology's Evolution and the Path of Measured Productivity Growth: Present and Future in the Mirror of the Past." In Brynjolfsson, Eric and Brian Kahin, eds. 2000. *Understanding the Digital Economy.* Cambridge, MA: The MIT Press.

David, Paul. 2001. "Productivity Growth Prospects and the New Economy in Historical Perspective." *Eib Papers* 6(1):41–62

DeLong, Bradford and Lawrence H. Summers. 2001. "The 'New Economy': Background, Historical Perspective, Questions, and Speculations." *Federal Reserve Bank of Kansas City Economic Review* 86(4):29–59.

Diewert, Irwin W. and Denis A. Lawrence. 2000. "Progress in Measuring the Price and Quantity of Capital." In Lawrence J. Lau, ed. *Econometrics and the Cost of Capital.* Cambridge, MA: The MIT Press.

Doms, Mark, 2004. "The Boom and Bust in Information Technology Investment." *Federal Reserve Bank of San Francisco Economic Review*: 19–34.

Doms, Mark. 2005. "Communications Equipment: What has Happened to Prices?" in Corrado, Carol A., John Haltiwanger, and Daniel Sichel, eds. *Measuring Capital in a New Economy.* Chicago, IL: University of Chicago Press.

Dulberger, Ellen R. 1989. "The Application of a Hedonic Model to a Quality Adjusted Price Index for Computer Processors." In Dale W. Jorgenson and Ralph Landau, eds. *Technology and Capital Formation*: Cambridge, MA: The MIT Press.

Dulberger, Ellen R. 1993. "Sources of Price Decline in Computer Processors: Selected Electronic Components." In Murray Foss, Marilyn Manser, and Allan Young, eds. *Price Measurements and Their Uses.* Chicago: University of Chicago Press for the National Bureau of Economic Research.

Easterly, William. 2001. *The Elusive Quest for Growth.* Cambridge MA: The MIT Press.

The Economist. 1997. "Assembling the New Economy." September 11.

The Economist. 2000. "To boldly go…" March 23.

The Economist. 2000. "A Thinker's Guide." March 30.

The Economist. 2000. "Productivity on Stilts." June 8.

The Economist. 2000. Performing Miracles." June 15.

The Economist. 2000. "Solving the Paradox." September 21.

The Economist. 2000. "Elementary, My Dear Watson." September 21.

The Economist. 2000. "Waiting for the New Economy." October 12.

The Economist. 2001. "The Great Chip Glut." August 11.

The Economist. 2001. "Productivity Growth (cont'd?)." September 6.

The Economist. 2003. "The New Geography of the IT Industry." July 17.

The Economist. 2003. "Overproductive and Underemployed." August 11.

The Economist. 2003. "Survey of the New Economy." September 11.

The Economist. 2003. "Between a Rock and a Hard Place." October 9.

The Economist. 2003. "Untangling the Local Loop." October 9.

The Economist. 2003. "Relocating the Back Office." December 11.

The Economist. 2004. "Innovative India." April 1.

The Economist. 2004. "Survey: A World of Work." November 11.

The Economist. 2005. "Moore's Law at 40." March 23.

The Economist. 2005. "How the Internet Killed the Phone Business." September 15.

Electronic News. 1999. "Sematech Adds 4 International Members." June 21.

Electronic News. 2005. "Samsung Faces $300M DoJ Fine for Price Fixing." October 13.

Ericson, R. and A. Pakes. 1995. "Markov-Perfect Industry Dynamics: A Framework for Empirical Work." *Review of Economic Studies* 62:53–82.

European Semiconductor Industry Association. 2005. *The European Semiconductor Industry 2005 Competitiveness Report.* Brussels, Belgium: European Semiconductor Industry Association. Accessed at http://www.eeca.org/pdf/final_comp_report.pdf.

Evans, Richard. 2002. "INSEE's Adoption of Market Intelligence Data for its Hedonic Computer Manufacturing Price Index." Presented at the Symposium on Hedonics at Statistics Netherlands. October 25.

Executive Office of the President. 2003. *The National Strategy to Secure Cyberspace: Cyberspace Threats and Vulnerabilities.* Washington, D.C.: Executive Office of the President.

Feenstra, Robert C., Marshall B. Reinsdorf, and Michael Harper. 2005. "Terms of Trade Gains and U.S. Productivity Growth." Paper prepared for NBER-CRIW Conference. July 25.

Fershtman, C. and A. Pakes. 2000. "A Dynamic Game with Collusion and Price Wars." *RAND Journal of Economics.* 31(2):207–236.

Fisher, Franklin M., John J. McGowan, and Joen E. Greenwood. 1983. *Folded, Spindled, and Multiplied: Economic Analysis and U.S. v. IBM.* Cambridge, MA: The MIT Press.

Flamm, Kenneth. 1988. *Creating the Computer.* Washington, D.C.: The Brookings Institution.

Flamm, Kenneth. 1993. "Measurement of DRAM Prices: Technology and Market Structure." In Murray F. Foss, Marilyn E. Manser, and Allan H. Young, eds. *Price Measurements and Their Uses.* Chicago: University of Chicago Press.

Flamm, Kenneth. 1996. *Mismanaged Trade? Strategic Policy and the Semiconductor Industry.* Washington, D.C.: The Brookings Institution.

Flamm, Kenneth. 1997. *More for Less: The Economic Impact of Semiconductors.* San Jose, CA: Semiconductor Industry Association.

Flamm. Kenneth. 2004. "Moore's Law and the Economics of Semiconductor Price Trends." In National Research Council. *Productivity and Cyclicality in Semiconductors: Trends, Implications, and Questions.* Dale W. Jorgenson and Charles W. Wessner, eds. Washington, D.C.: The National Academies Press.

Kenneth Flamm, 2005. "The Coming IT Slowdown: Technological Roots and Economic Implications." Working Paper, LBJ School of Public Policy, University of Texas.

Frankel, Jeffrey and Peter Orsag, eds. 2002. *American Economic Policy in the 1990s.* Cambridge, MA: The MIT Press.

Fransman, M. 1992. *The Market and Beyond: Cooperation and Competition in Information Technology Development in the Japanese System.* Cambridge, UK: Cambridge University Press.

Gandal, Neil. 1994. "Hedonic Price Indexes for Spreadsheets and an Empirical Test for Network Externalities." *RAND Journal of Economics* 25.

Gordon, Robert J. 1989. "The Postwar Evolution of Computer Prices." In Dale W. Jorgenson and Ralph Landau, eds. *Technology and Capital Formation.* Cambridge, MA: The MIT Press.

Gordon, Robert J. 1999. "Has the 'New Economy' Rendered the Productivity Slowdown Obsolete?" Paper presented at the Federal Reserve Bank of Chicago, June 9.

Gordon, Robert J., 2000. "Does the 'New Economy' Measure up to the Great Inventions of the Past?" *Journal of Economic Perspectives* 14(4). Northwestern University Working Paper.

Gordon, Robert J. 2000. "Interpreting the 'One Big Wave' in U.S. Long-Term Productivity Growth." NBER Working Paper 7752.

Gordon, Robert J. 2004. "Innovation and Future Productivity Growth: Does Supply Create its own Demand?" In Peter Cornelius, ed. *The Global Competitiveness Report 2002-3.* New York: Oxford University Press.

Gosling, James, Bill Joy, and Guy Steele. 1996. *The Java (TM) Language Specification.* New York: Addison-Wesley.

Griffith, P. 1993. "Science and the Public Interest." *The Bridge*. Washington, D.C.: National Academy of Engineering. (Fall):16.

Griliches, Zvi. 1960. "Measuring Inputs in Agriculture: A Critical Survey." *Journal of Farm Economics* 40(5):1398–1427.

Griliches, Zvi. 1961. "Hedonic Price Indexes for Automobiles: An Econometric Analysis of Quality Change." In George Stigler, ed. *The Price Statistics of the Federal Government*. New York: Columbia University Press.

Griliches, Zvi. 1994. "Productivity, R&D, and the Data Constraint." *American Economic Review* 94(2):1–23.

Grindley, P., D. C. Mowery, and B. Silverman. 1994. "SEMATECH and Collaborative Research: Lessons in the Design of a High-Technology Consortia." *Journal of Policy Analysis and Management* 13.

Grossman, Gene and Elhannan Helpman. 1993. *Innovation and Growth in the Global Economy*. Cambridge, MA: The MIT Press.

Gowrisankaran, G. 1998. "Issues and Prospects for Payment System Deregulation." Working Paper. University of Minnesota.

Hagel, J. and A. G. Armstrong. 1997. *Net Gain*. Cambridge, MA: Harvard Business School Press.

Halstead, Maurice H. 1977. *Elements of Software Science*. New York: Elsevier North Holland.

Handler, Philip. 1970. *Biology and the Future of Man*. London, UK: Oxford University Press.

Harhoff, Dietmar and Dietmar Moch. 1997. "Price Indexes for PC Database Software and the Value of Code Compatibility." *Research Policy* 24(4-5):509–520.

Hazlett, Thomas W. and George Bittlingmayer. 2003. "The Political Economy of Cable 'Open Access.'" *Stanford Technology Law Review* 4.

Holdway, Michael. 2001. "Quality-Adjusting Computer Prices in the Producer Price Index: An Overview." Working Paper. Washington, D.C.: Bureau of Labor Statistics.

Holdway, Michael. 2002 "Confronting the Challenge of Estimating Constant Quality Price Indexes for Telecommunications Equipment in the Producer Price Index." Working Paper. Washington, D.C.: Bureau of Labor Statistics.

Hornstein, Andreas and Per Krusell. 2000. "The IT Revolution: Is It Evident in the Productivity Numbers?" *Federal Reserve Bank of Richmond Economic Quarterly* 86(4).

Horrigan, John Brendan. 2005. "Broadband Adoption at Home in the United States: Growing but Slowing." Washington, D.C.: Pew Internet and American Life Project.

Howell, Thomas. 2003. "Competing Programs: Government Support for Microelectronics." In National Research Council. *Securing the Future: Regional and National Programs to Support the Semiconductor Industry*. Charles W. Wessner, ed. Washington, D.C.: The National Academies Press.

Information Technology Association of America. 2004. "The Impact of Offshore IT Software and Services Outsourcing on the U.S. Economy and the IT Industry." Prepared by Global Insight.

Isenberg, David. 1997. "Rise of the Stupid Network" *Computer Telephony* (August):16–26.

Ishida, Haruhisa. 1972. "On the Origin of the Gibson Mix." *Journal of the Information Processing Society of Japan* 13(May):333–334 (in Japanese).

Jorgenson, Dale W. 2001. *Economic Growth in the Information Age*. Cambridge, MA: The MIT Press.

Jorgenson, Dale W. 2001. "Information Technology and the U.S. Economy." *American Economic Review* 91(1).

Jorgenson, Dale W. 2002. *Economic Growth in the Information Age-Volume 3*. Cambridge, MA and London, UK: The MIT Press.

Jorgenson, Dale W. 2002. "The Promise of Growth in the Information Age." The Conference Board Annual Essay.

Jorgenson, Dale W. 2004. "Information Technology and the World Economy." Leon Kozminsky Academy Distinguished Lecture. May 14.

Jorgenson, Dale W., Mun S. Ho, and Kevin J. Stiroh. 2002. "Projecting Productivity Growth: Lessons from the U.S. Growth Resurgence." *Federal Reserve Bank of Atlanta Economic Review* 87(3):1–13.

Jorgenson, Dale W., Mun S. Ho, and Kevin J. Stiroh. 2004. "Will the U.S. Productivity Resurgence Continue?" *Federal Reserve Bank of New York Current Issues in Economics and Finance* 10(13):1–7.

Jorgenson, Dale W., Mun S. Ho, and Kevin J. Stiroh. 2005. "Growth of U.S. Industries and Investments in Information Technology and Higher Education." In Corrado, Carol A., John Haltiwanger, and Daniel Sichel, eds. *Measuring Capital in a New Economy*. Chicago, IL: University of Chicago Press.

Jorgenson, Dale W., J. Steven Landefeld, and William Nordhaus, eds. 2006. *A New Architecture for the U.S. National Accounts*. Chicago, IL: University of Chicago Press.

Jorgenson, Dale W. and Kevin J. Stiroh. 1999. "Productivity Growth: Current Recovery and Longer-term Trends." *American Economic Review*. 89(2):109–115.

Jorgenson, Dale W. and Kevin J. Stiroh. 2002. "Raising the Speed Limit: U.S. Economic Growth in the Information Age." In National Research Council. *Measuring and Sustaining the New Economy*. Dale W. Jorgenson and Charles W. Wessner, eds. Washington, D.C.: National Academy Press.

Jorgenson, Dale W. and Eric Yip. 2000. "Whatever Happened to Productivity Growth?" In *New Developments in Productivity Analysis,* Charles R. Hulten, Edwin R. Rean and Michael J. Harper, eds. Boston, MA: National Bureau of Economic Research.

Jovanovic, Boyan and Peter L. Rousseau. 2002. "Moore's Law and Learning-by-Doing." *Review of Economic Dynamics* 5:346–375.

Kessler, Michelle. 2002. "Computer Majors Down Amid Tech Bust." *USA Today*. October 8.

Knight, Kenneth E. 1966. "Changes in Computer Performance: A Historical View." *Datamation* (September):40–54.

Knight, Kenneth E. 1970. "Application of Technological Forecasting to the Computer Industry." In James R. Bright and Milton E.F. Schieman, *A Guide to Practical Technological Forecasting*. Englewood Cliffs, NJ: Prentice-Hall.

Knight, Kenneth E. 1985. "A Functional and Structural Measure of Technology." *Technological Forecasting and Technical Change* 27(May):107–127.

Koskimäki, Timo and Yrjö Vartia. 2001. "Beyond matched pairs and Griliches-type hedonic methods for controlling quality changes in CPI sub-indices." Presented at Sixth Meeting of the International Working Group on Price Indices, sponsored by the Australian Bureau of Statistics, April.

Kuan, Jennifer. 2005. "Open Source Software as Lead User's Make or Buy Decision: A Study of Open and Closed Source Quality." Palo Alto, CA: Stanford Institute for Economic Policy Research.

Landefeld, J. Steven and Bruce Grimm. 2000. "A Note on the Impact of Hedonics and Computers on Real GDP." *Survey of Current Businesses* 80(12):17–22.

Landefeld, J. Steven and Barbara M. Fraumeni. 2001. "Measuring the New Economy." *Survey of Current Business* 81(3):23–40.

Landefeld, J. Steven and Robert P. Parker. 1997. "BEA's Chain Indexes, Time Series, and Measures of Long-term Growth." *Survey of Current Business* 77(5):58–68.

Levine, Jordan. 2002. "U.S. Producer Price Index for Pre-Packaged Software." Presented at the 17th Voorburg Group Meeting. Nantes, France. September.

Levy, David and Steve Welzer. 1985. "An Unintended Consequence of Antitrust Policy: The Effect of the IBM Suit on Pricing Policy." Unpublished Paper. Rutgers University Department of Economics.

Levy, David L. 2005. "The New Global Political Economy." *Journal of Management Studies* 42(3):685.

Lim, Poh Ping and Richard McKenzie. 2002. "Hedonic Price Analysis for Personal Computers in Australia: An Alternative Approach to Quality Adjustments in the Australian Price Indexes." Paper presented at ZEW conference. Mannheim, Germany. April.

Litan, Robert E. and Roger G. Noll. 2003. "The Uncertain Future of the Telecommunications Industry." Brookings Working Paper. Washington D.C.: The Brookings Institution.

Litan, Robert E. and Alice M. Rivlin, 2000. "The Economy and the Internet: What Lies Ahead?" Washington, D.C.: Internet Policy Institute. Accessed at http://www.intenetpolicy.org/briefing/litan_rivlin.html.

Macher, Jeffrey T., David C. Mowery, and David A. Hodges. 1999. "Semiconductors." In National Research Council. *U.S. Industry in 2000: Studies in Competitive Performance*. David C. Mowery, ed. Washington, D.C.: National Academy Press.

Maney, Kevin. 2003. "Music industry doesn't know what else to do as it lashes out at file-sharing." *USA Today*. September 9.

Mann, Catherine. 2003. "Globalization of IT Services and White Collar Jobs: The Next Wave of Productivity Growth." *International Economics Policy Briefs* PB03-11. December.

Mann, Catherine. 2004. "The U.S. Current Account, New Economy Services, and Implications for Sustainability." *Review of International Economics* 12(2):262–276.

Mann, Catherine L. Forthcoming. *High Technology and the Globalization of America*.

Martin, Brookes and Zaki Wahhaj. 2000. "The Shocking Economic Impact of B2B." *Global Economic Paper, 37*. Goldman Sachs. February 3.

Maxwell, Kim. 1998. *Residential Broadband: An Insider's Guide to the Battle for the Last Mile*. Hoboken, NJ: John Wiley & Sons.

McKinsey Global Institute. 2001. *U.S. Productivity Growth 1995-2000: Understanding the Contribution of Information Technology Relative to Other Factors*. Washington, D.C.: McKinsey & Company.

McKinsey Global Institute. 2003. *Off-shoring: Is it a Win-Win Game?* San Francisco, CA: McKinsey & Company.

McKinsey Global Institute. 2005. *The Emerging Global Labor Market*. Washington D.C.: McKinsey & Company.

Michaels, Robert. 1979. "Hedonic Prices and the Structure of the Digital Computer Industry." *The Journal of Industrial Economics* 27 (March):263–275.

Moch, Dietmar. 2001. "Price Indices for Information and Communication Technology Industries: An Application to the German PC Market." Center for European Economic Research (ZEW) Discussion Paper No. 01-20. Mannheim, Germany: ZEW.

Moore, Gordon E. 1965. "Cramming More Components onto Integrated Circuits." *Electronics* 38(8) April.

Moore, Gordon E. 1975. "Progress in Digital Integrated Circuits." *Proceedings of the 1975 International Electron Devices Meeting* 11–13.

Moore, Gordon E. 1997. "The Continuing Silicon Technology Evolution Inside the PC Platform." *Intel Developer Update*. Issue 2.

Moore, Gordon E. 2003. "No Exponential if Forever . . . but We can Delay Forever." Santa Clara, CA: Intel Corporation.

Moylan, Carol. 2001. "Estimation of Software in the U.S. National Income and Product Accounts: New Developments." Paris, France: Organisation for Economic Co-operation and Development.

National Academy of Sciences, National Academy of Engineering, Institute of Medicine. 1993. *Science, Technology and the Federal Government. National Goals for a New Era*. Washington, D.C.: National Academy Press.

National Advisory Committee on Semiconductors. 1992. *A National Strategy for Semiconductors: An Agenda for the President, the Congress, and the Industry*. Washington, D.C.: National Advisory Committee on Semiconductors.

National Institute of Standards and Technology. 2002. "The Economic Impacts of Inadequate Infrastructure for Software Testing." Planning Report 02-3. Gaithersburg, MD: U.S. Department of Commerce.

National Research Council. 1995. *Standards, Conformity Assessment, and Trade into the 21st Century.* Washington, D.C.: National Academy Press.

National Research Council. 1996. *Conflict and Cooperation in National Competition for High-Technology Industry.* Washington, D.C.: National Academy Press.

National Research Council. 1999. *The Advanced Technology Program: Challenges and Opportunities.* Charles W. Wessner, ed. Washington, D.C.: National Academy Press.

National Research Council. 1999. *Industry-Laboratory Partnerships: A Review of the Sandia Science and Technology Park Initiative.* Charles W. Wessner, ed. Washington, D.C.: National Academy Press.

National Research Council. 1999. *New Vistas in Transatlantic Science and Technology Cooperation.* Charles W. Wessner, ed. Washington, D.C.: National Academy Press.

National Research Council. 1999. *The Small Business Innovation Research Program: Challenges and Opportunities.* Charles W. Wessner, ed. Washington, D.C.: National Academy Press.

National Research Council. 2001. *The Advanced Technology Program: Assessing Outcomes.* Charles W. Wessner, ed. Washington, D.C.: National Academy Press.

National Research Council. 2001. *Capitalizing on New Needs and New Opportunities: Government-Industry Partnerships in Biotechnology and Information Technologies.* Charles W. Wessner, ed. Washington, D.C.: National Academy Press.

National Research Council. 2001. *A Review of the New Initiatives at the NASA Ames Research Center.* Charles W. Wessner, ed. Washington, D.C.: National Academy Press.

National Research Council. 2001. *Trends in Federal Support of Research and Graduate Education.* Stephen A. Merrill, ed. Washington, D.C.: National Academy Press.

National Research Council. 2002. *Measuring and Sustaining the New Economy.* Dale W. Jorgenson and Charles W. Wessner, eds. Washington, D.C.: National Academy Press.

National Research Council. 2002. *Partnerships for Solid-State Lighting.* Charles W. Wessner, ed. Washington, D.C.: National Academy Press.

National Research Council. 2003. *Government-Industry Partnerships for the Development of New Technologies: Summary Report.* Charles W. Wessner, ed. Washington, D.C.: The National Academies Press.

National Research Council. 2003. *Securing the Future: Regional and National Programs to Support the Semiconductor Industry.* Charles W. Wessner, ed. Washington, D.C.: The National Academies Press.

National Research Council. 2004. *Productivity and Cyclicality in Semiconductors: Trends, Implications, and Questions.* Dale W. Jorgenson and Charles W. Wessner, eds., Washington, D.C.: The National Academies Press.

National Research Council. 2004. *The Small Business Innovation Research Program: Program Diversity and Assessment Challenges.* Charles W. Wessner, ed. Washington, D.C.: The National Academies Press.

National Research Council, 2005. *Deconstructing the Computer.* Dale W. Jorgenson and Charles W. Wessner, eds. Washington, D.C.: The National Academies Press.

National Research Council. 2005. *Getting Up to Speed: The Future of Supercomputing.* Washington, D.C.: The National Academies Press.

National Research Council. 2005. *Policy Implications of International Graduate Students and Postdoctoral Scholars in the United States.* Washington, D.C.: The National Academies Press.

National Research Council. 2005. *Rising Above the Gathering Storm: Energizing and Employing America for a Brighter Economic Future.* Washington, D.C.: The National Academies Press.

National Research Council. 2006. *Software, Growth, and the Future of the U.S. Economy.* Dale W. Jorgenson and Charles W. Wessner, eds. Washington, D.C.: The National Academies Press.

National Research Council. Forthcoming. *Enhancing Productivity Growth in the Information Age: Measuring and Sustaining the New Economy.* Charles W. Wessner, ed. Washington, D.C.: The National Academies Press.

Nelson, Richard, ed. 1993. *National Innovation Systems.* New York: Oxford University Press.

Nelson, R. A., T. L. Tanguay, and C. C. Patterson. 1994. "A Quality-adjusted Price Index for Personal Computers." *Journal of Business and Economics Statistics* 12(1):23–31.

The New York Times. 2003. "Good Economy. Bad Job Market. Huh?" September 14.

The New York Times. 2004. "Financial Firms Hasten Their Move to Outsourcing." August 14.

Nikkei Microdevices. 2001. "From Stagnation to Growth: The Push to Strengthen Design." January.

Nordhaus, William D. 2002. "Productivity Growth and the New Economy" *Brookings Papers on Economic Activity* 2:211–44.

Nordhaus, William D. 2002. "The Progress of Computing." New Haven, CT: Yale University. March 4.

Okamoto, Masato and Tomohiko Sato. 2001. "Comparison of Hedonic Method and Matched Models Method Using Scanner Data: The Case of PCs, TVs and Digital Cameras." Presented at Sixth Meeting of the International Working Group on Price Indices, sponsored by the Australian Bureau of Statistics. April.

Oliner, Stephen. 2000. The Resurgence of Growth in the Late 1990s: Is Information Technology the Story?" *Journal of Economic Perspectives* 14(4):3–22.

Oliner, Stephen. 2002. "Information Technology and Productivity: Where are We Now and Where are We Going?" *Federal Bank of Atlanta Economic Review* 87(3):15–44.

O'Mahoney, Mary and Bart van Ark. 2003. *EU Productivity and Competitiveness: An Industry Perspective: Can Europe Resume the Catching-up Process?* Luxembourg: Office for Official Publications of the European Communities.

Organisation for Economic Co-operation and Development. 2000. *Is There a New Economy? A First Report on the OECD Growth Project.* Paris, France: Organisation for Economic Co-operation and Development.

Organisation for Economic Co-operation and Development. 2002. *Information Technology Outlook 2002—The Software Sector.* Paris, France: Organisation for Economic Co-operation and Development. p. 105.

Organisation for Economic Co-operation and Development. 2003. *ICT and Economic Growth.* Paris, France: Organisation for Economic Co-operation and Development.

Organisation for Economic Co-operation and Development. 2003. Report of the OECD Task Force on Software Measurement in the National Accounts. Statistics Working Paper 2003/1. Paris, France: Organisation for Economic Co-operation and Development.

PC World Magazine. 2003. "20 Years of Hardware." March.

Paganetto, Luigi, ed. 2004. *Knowledge Economy, Information Technologies, and Growth.* Burlington, VT: Ashgate.

Pakes, G. E. 1966. *Physics Survey and Outlook.* National Academy of Sciences. Washington D.C.: National Academy Press.

Pakes, Ariel. 2001. "A Reconsideration of Hedonic Price Indices with an Application to PCs." Harvard University. November.

Parker, Robert P., and Bruce Grimm. 2000. "Recognition of Business and Government Expenditures for Software as Investment: Methodology and Quantitative Impacts, 1959-98." Washington, D.C.: Bureau of Economic Analysis.

Porter, Michael. 2004. "Building the Microeconomic Foundations of Prosperity: Findings from the Business Competitiveness Index." In X Sala-i-Martin, ed. *The Global Competitiveness Report 2003-2004.* New York: Oxford University Press.

Poulsen, Kevin. 2004. "Software Bug Contributed to Blackout." *Security Focus.* February 11.

Pritchard, Stephen. 2003. "Munich Makes the Move." *Financial Times.* October 15.

Prud'homme, Marc and Kam Yu. 2002. "A Price Index for Computer Software Using Scanner Data." Unpublished working paper, Prices Division, Statistics Canada. Ottawa, ON.

Raduchel, William. 2006. "The Economics of Software." In National Research Council, *Software, Growth, and the Future of the U.S. Economy*. Dale W. Jorgenson and Charles W. Wessner, eds. Washington, D.C.: The National Academies Press.

Rao, H. Raghaw and Brian D. Lynch. 1993. "Hedonic Price Analysis of Workstation Attributes." *Communications of the Association for Computing Machinery (ACM)* 36(12):94–103.

Robertson, Jack. 1998. "Die Shrinks Now Causing Logic Chip Glut." *Semiconductor Business News*. October 15.

Ruttan, Vernon. 2001. *Technology, Growth, and Development*. New York: Oxford University Press.

Samuelson, Paul. 2004. "Why Ricardo and Mill Rebut and Confirm Arguments of Mainstream Economists Supporting Globalization." *Journal of Economic Perspectives* 18(3).

Schaller, Robert R. 1999. "Technology Roadmaps: Implications for Innovation, Strategy, and Policy." Ph.D. Dissertation Proposal, Institute for Public Policy, George Mason University.

Schaller, Robert R. 2002. "Moore's Law: Past, Present, and Future." Accessed at <http://www.njtu.edu.cn/depart/xydzxx/ec/spectrum/moore/mlaw.html> on July 2002.

Schlender, Brent. 1999. "The Edison of the Internet." *Fortune* February 15.

Semiconductor Industry Association. 2004. *International Technology Roadmap for Semiconductors: Update*. Accessed at <http://www.itrs.net/Common/2004Update/2004Update.htm>.

Semiconductor Industry Association. 2005. *International Technology Roadmap for Semiconductors*. Accessed at <http://www.itrs.net/Common/2005ITRS/Home2005.htm>.

Semiconductors International. 2000. "Sematech Forms International Sematech." March.

Shapiro, Carl and Hal R. Varian. 1999. *Information Rules*. Boston, MA: Harvard Business School Press.

Sharpe, William F. 1969. *The Economics of the Computer*. New York, NY and London, UK: Columbia University Press.

Sichel, Daniel, E. 1997. *Computer Revolution: An Economic Perspective*. Washington, D.C.: The Brookings Institution.

Siebert, Horst. 2002. *Economic Policy Issues of the New Economy*. Heidelberg, Germany and New York: Springer.

Sigurdson, Jon. 1986. *Industry and State Partnership in Japan: The Very Large Scale Integrated Circuits (VLSI) Project*. Lund, Sweden: Research Policy Institute.

Sirgudson, Jon. 2004. "VSLI Revisited—Revival in Japan." Working Paper No. 191. Tokyo, Japan: Institute of Innovation Research of Hitotsubashi University.

Soete, Luc. 2001. "The New Economy: A European Perspective." In Daniele Archibugi and Bengt-Ake Lundvall, eds. *The Globalizing Learning Economy*. Oxford, UK and New York: Oxford University Press.

Solow, Robert M. 1987. "We'd Better Watch Out." *New York Times Book Review*. July 12.

Solow, Robert M., Michael Dertouzos, and Richard Lester. 1989. *Made in America*. Cambridge, MA: The MIT Press.

Spencer, W. J. and P. Grindley 1993. "SEMATECH After Five Years: High Technology Consortia and U.S. Competitiveness." *California Management Review*. 35.

Spencer, W. J. and T. E. Seidel. 2004. "National Technology Roadmaps: The US Semiconductor Experience." In National Research Council. Productivity and Cyclicality in Semiconductors: Trends, Implications, and Questions. Charles W. Wessner, ed. Washington, D.C.: The National Academies Press.

Statistics Finland. 2000. "Measuring the Price Development of Personal Computers in the Consumer Price Index." Paper for the Meeting of the International Hedonic Price Indexes Project. Paris, France. September 27.

Stiroh, Kevin J. 2001. "Investing in Information Technology: Productivity Payoffs for U.S. Industries. *Federal Reserve Bank of New York Current Issues in Economics and Finance* 7(6):1–6.

Stiroh, Kevin J. 2002. "Are ICT Spillovers Driving the New Economy?" *Review of Income and Wealth* 48(1).

Stiroh, Kevin J. 2002 "Information Technology and the U.S. Productivity Revival: What do the Industry Data Say?" *American Economic Review* 92(5):1559–1576.

Stiroh, Kevin J. 2002. "Measuring Information Technology and Productivity in the New Economy" *World Economics* 3(1):43–58.

Stoneman, Paul. 1976. Technological Diffusion and the Computer Revolution: The U.K. Experience. Cambridge, UK: Cambridge University Press.

Temple, Jonathan. 2002. "The Assessment: The New Economy." *Oxford Review of Economic Policy* 18(3):241–264.

Triplett, Jack E. 1985. "Measuring Technological Change with Characteristics-Space Techniques." *Technological Forecasting and Social Change* 27:283–307.

Triplett, Jack E. 1989. "Price and Technological Change in a Capital Good: A Survey of Research on Computers." In Dale W. Jorgenson and Ralph Landau, eds. *Technology and Capital Formation.* Cambridge, MA: The MIT Press.

Triplett, Jack E. 1996. "High-Tech Productivity and Hedonic Price Indexes." in Organisation for Economic Co-operation and Development. *Industry Productivity.* Paris, France: Organisation for Economic Co-operation and Development.

Triplett, Jack E. 1997. "The Solow Productivity Paradox: What do Computers do to Productivity?" Paper presented at the Conference on Service Sector Productivity and the Productivity Paradox, Ottawa, ON. April 11–12.

Triplettt, Jack E. 1999. "Did the U.S. have a New Economy?" Paper presented for the Association de Compatible Nationale 9th Conference on National Accounting.

Triplett, Jack E. 1999. "Economic Statistics, the New Economy, and the Productivity Slowdown." *Business Economics.* January.

Triplett, Jack E. 1999. "The Solow Productivity Paradox: What do Computers do to Productivity?" *Canadian Journal of Economics* 32(2):309–334.

Triplett, Jack E. and Barry Bosworth. 2002. "Baumol's Disease Has Been Cured: IT and Multifactor Productivity in U.S. Service Industries." Presented at the Brookings Workshop "Services Industry Productivity: New Estimates and New Problems." March 14. Accessible at <http://www.brook.edu/dybdocroot/es/research/projects/productivity/workshops/20020517.htm>.

United States Congress. 2001. "Information Technology and the New Economy." Washington, D.C.: Joint Economic Committee Study.

van Ark, Bart, Robert Inklaar, and Robert H. McGuckin. 2002. "Changing Gear, Productivity, ICT and Service Industries: Europe and the United States." Brookings Seminar Paper on Productivity in Services. Washington, D.C.: The Brookings Institution.

van Mulligen, Peter Hein. 2002. "Alternative Price Indices for Computers in the Netherlands Using Scanner Data." Prepared for the 27th General Conference of the International Association for Research in Income and Wealth, Djurhamn, Sweden.

van Welsum, Desirée and Xavier Reif. 2005. "Potential Off-shoring: Evidence from Selected OECD Countries." Paris, France: Organisation for Economic Co-operation and Development.

Vatter, Harold G. and John F. Walker. 2001. "Did the 1990s Inaugurate a New Economy?" *Challenge!* 44(1):90–116.

The Wall Street Journal. 2003. "Searching for Computing's Kilowatt." July 17.

The Wall Street Journal. 2004. "Outsourcing May Create U.S. Jobs." March 30.

The Washington Post. 2004. "Election Campaign Hit More Sour Notes." February 22.

Whelan, Karl. 2002. "Computers, Obsolescence, and Productivity." *Review of Economics and Statistics* 84(4):445–462.

Wolff, Alan Wm., Thomas R. Howell, Brent L. Bartlett, and R. Michael Gadbaw, eds. 1992. *Conflict Among Nations: Trade Policies in the 1990s.* San Francisco, CA: Westview Press.

World Semiconductor Trade Statistics. 2000. *Annual Consumption Survey.*

Wyckoff, Andrew W. 1995. "The Impact of Computer Prices on International Comparisons of Labour Productivity." *Economics of Innovation and New Technology* 3:277–293.